SCHAUM'S *Easy* OUTLINES

GENETICS

Other Books in Schaum's Easy Outline Series Include:

Schaum's Easy Outline: College Mathematics
Schaum's Easy Outline: College Algebra
Schaum's Easy Outline: Elementary Algebra
Schaum's Easy Outline: Calculus
Schaum's Easy Outline: Mathematical Handbook of Formulas and Tables
Schaum's Easy Outline: Geometry
Schaum's Easy Outline: Precalculus
Schaum's Easy Outline: Trigonometry
Schaum's Easy Outline: Probability and Statistics
Schaum's Easy Outline: Statistics
Schaum's Easy Outline: Principles of Accounting
Schaum's Easy Outline: Biology
Schaum's Easy Outline: College Chemistry
Schaum's Easy Outline: Human Anatomy and Physiology
Schaum's Easy Outline: Organic Chemistry
Schaum's Easy Outline: Physics
Schaum's Easy Outline: Programming with C++
Schaum's Easy Outline: Programming with Java
Schaum's Easy Outline: Basic Electricity
Schaum's Easy Outline: French
Schaum's Easy Outline: German
Schaum's Easy Outline: Spanish
Schaum's Easy Outline: Writing and Grammar

SCHAUM'S *Easy* OUTLINES

GENETICS

BASED ON SCHAUM'S
Outline of Genetics, Third Edition

BY WILLIAM D. STANSFIELD

ABRIDGEMENT EDITOR
GEORGE J. HADEMENOS

SCHAUM'S OUTLINE SERIES

McGRAW-HILL

New York Chicago San Francisco Lisbon London Madrid
Mexico City Milan New Delhi San Juan
Seoul Singapore Sydney Toronto

WILLIAM D. STANSFIELD is Professor Emeritus of Biological Sciences at California Polytechnic State University where he taught for twenty-nine years. He has a B.S. degree in agriculture, an M.S. in education, and a Ph.D. in genetics from the University of California at Davis. He has published research in immunogenetics, twinning, and mouse genetics, and he has written university level textbooks in evolution and serology/immunology as well as a dictionary of genetics.

GEORGE J. HADEMENOS has taught at the University of Dallas and done research at the University of Massachusetts Medical Center and the University of California at Los Angeles. He holds a B.S. degree from Angelo State University and both M.S. and Ph.D. degrees from the University of Texas at Dallas. He is the author of several books in the Schaum's Outline and Schaum's Easy Outline series.

3 4 5 6 7 8 9 0 DOC DOC 0 9 8 7 6

ISBN 0-07-138317-4

McGraw-Hill

A Division of The ***McGraw-Hill*** *Companies*

Contents

Chapter 1
The Physical Basis of Heredity

In This Chapter:

- ✔ *Genetics*
- ✔ *Cells*
- ✔ *Chromosomes*
- ✔ *Cell Division*
- ✔ *Mendel's Laws*
- ✔ *Gametogenesis*
- ✔ *Life Cycles*

Genetics

Genetics is that branch of biology concerned with heredity and variation. The hereditary units that are transmitted from one generation to the next (inherited) are called **genes**. The genes reside in a long molecule called **deoxyribonucleic acid** (DNA). The DNA, in conjunction with a protein matrix, forms **nucleoprotein** and becomes organized into structures with distinctive staining properties called **chromosomes** found in the nucleus of the cell. The behavior of genes is thus paralleled in many ways by the behavior of the chromosomes of which they are a part.

A gene contains coded information for the production of proteins. DNA is normally a stable molecule with the capacity for self-replication. On rare occasions, a change may occur spontaneously in some part of DNA. This change, called a **mutation**, alters the coded instructions and may result in a defective protein or in the cessation of protein synthesis. The net result of a mutation is often seen as a change in the physical appearance of the individual or a change in some other measurable attribute of the organism called a **character** or **trait**. Through the process of mutation, a gene may be changed into two or more alternative forms called **allelomorphs** or **alleles**.

Each gene occupies a specific position on a chromosome, called the gene **locus** (**loci**, plural). All allelic forms of a gene, therefore, are found at corresponding positions on genetically similar (**homologous**) chromosomes. The word "locus" is sometimes used interchangeably for "gene." When the science of genetics was in its infancy, the gene was thought to behave as a unit particle. These particles were believed to be arranged on the chromosome like beads on a string. All the genes on a chromosome are said to be **linked** to one another and belong to the same **linkage group**. Wherever the chromosome goes, it carries all of the genes in its linkage group with it. Linked genes are not transmitted independently of one another, but genes in different linkage groups (on different chromosomes) are transmitted independently of one another.

Cells

The smallest unit of life is the **cell**. Each living thing is composed of one or more cells. The most primitive cells alive today are the bacteria. They, like the presumed first forms of life, do not possess a **nucleus**. The nucleus is a membrane-bound compartment, isolating the genetic material from the rest of the cell (**cytoplasm**). Bacteria, therefore, belong to a group of organisms called **procaryotes** (literally, "before a nucleus" had evolved; also spelled prokaryotes). All other kinds of cells that have a nucleus (including fungi, plants, and animals) are referred to as **eucaryotes** (literally, "truly nucleated"; also spelled eukaryotes).

The cells of a multicellular organism seldom look alike or carry out identical tasks. The cells are differentiated to perform specific functions (sometimes referred to as a "division of labor"). A neuron is specialized to conduct nerve impulses, a muscle cell contracts, a red blood cell carries oxygen, and so on. Thus, there is no such thing as a typical cell type. Figure 1-1 is a composite diagram of an animal cell showing common subcellular structures that are found in all or most cell types.

Any subcellular structure that has a characteristic morphology and function is considered to be an **organelle**. Some of the organelles (such as the nucleus and mitochondria) are membrane-bound; others (such as the ribosomes and centrioles) are not enclosed by a membrane. Most organelles and other cell parts are too small to be seen with the light microscope, but they can be studied with the electron microscope. The characteristics of organelles and other parts of eucaryotic cells are outlines in Table 1.1.

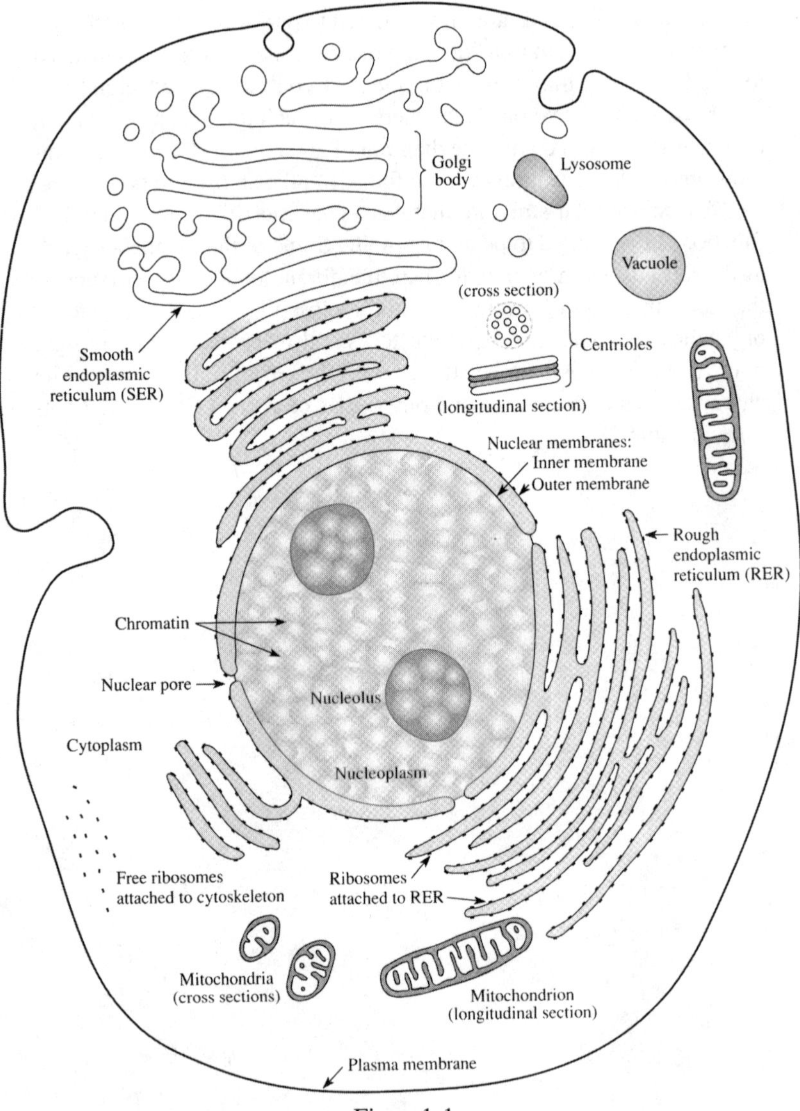
Golgi body
Lysosome
Vacuole
(cross section)
Centrioles
(longitudinal section)
Smooth endoplasmic reticulum (SER)
Nuclear membranes:
Inner membrane
Outer membrane
Rough endoplasmic reticulum (RER)
Chromatin
Nuclear pore
Nucleolus
Cytoplasm
Nucleoplasm
Free ribosomes attached to cytoskeleton
Ribosomes attached to RER
Mitochondria (cross sections)
Mitochondrion (longitudinal section)
Plasma membrane

Figure1-1

Table 1.1. Characteristics of Eucaryotic Cellular Structures

Cell Structures	Characteristics
Extracellular structures	A cell wall surrounding the plasma membrane gives strength and rigidity to the cell and is composed primarily of cellulose in plants (peptidoglycans in bacterial ''envelopes''); animal cells are not supported by cell walls; slime capsules composed of polysaccharides or glycoproteins coat the cell walls of some bacterial and algal cells
Plasma membrane	Lipid bilayer through which extracellular substances (e.g., nutrients, water) enter the cell and waste substances or secretions exit the cell; passage of substances may require expenditure of energy (active transport) or may be passive (diffusion)
Nucleus	Master control of cellular functions via its genetic material (DNA)
Nuclear membrane	Double membrane controlling the movement of materials between the nucleus and cytoplasm; contains pores that communicate with the ER
Chromatin	Nucleoprotein component of chromosomes (seen clearly only during nuclear division when the chromatin is highly condensed); only the DNA component is hereditary material
Nucleolus	Site(s) on chromatin where ribosomal RNA (rRNA) is synthesized; disappears from light microscope during cellular replication
Nucleoplasm	Nonchromatin components of the nucleus containing materials for building DNA and messenger RNA (mRNA molecules serve as intermediates between nucleus and cytoplasm)
Cytoplasm	Contains multiple structural and enzymatic systems (e.g., glycolysis and protein synthesis) that provide energy to the cell; executes the genetic instructions from the nucleus

Table 1.1. (continued)

Cell Structures	Characteristics
Ribosome	Site of protein synthesis; consists of three molecular weight classes of ribosomal RNA molecules and about 50 different proteins
Endoplasmic reticulum	Internal membrane system (designated ER); rough endoplasmic reticulum (RER) is studded with ribosomes and modifies polypeptide chains into mature proteins (e.g., by glycosylation); smooth endoplasmic reticulum (SER) is free of ribosomes and is the site of lipid synthesis
Mitochondria	Production of adenosine triphosphate (ATP) through the Krebs cycle and electron transport chain; beta oxidation of long-chain fatty acids; ATP is the main source of energy to power biochemical reactions
Plastid	Plant structure for storage of starch, pigments, and other cellular products; photosynthesis occurs in chloroplasts
Golgi body (apparatus)	Sometimes called dictyosome in plants; membranes where sugars, phosphate, sulfate, or fatty acids are added to certain proteins; as membranes bud from the Golgi system they are marked for shipment in transport vesicles to arrive at specific sites (e.g., plasma membrane, lysosome)
Lysosome	Sac of digestive enzymes in all eucaryotic cells that aid in intracellular digestion of bacteria and other foreign bodies; may cause cell destruction if ruptured
Vacuole	Membrane-bound storage deposit for water and metabolic products (e.g., amino acids, sugars); plant cells often have a large central vacuole that (when filled with fluid to create turgor pressure) makes the cell turgid
Centrioles	Form poles of the spindle apparatus during cell divisions; capable of being replicated after each cell division; rarely present in plants

Cell Structures	Characteristics
Cytoskeleton	Contributes to shape, division, and motility of the cell and the ability to move and arrange its components; consists of microtubules of the protein tubulin (as in the spindle fibers responsible for chromosomal movements during nuclear division or in flagella and cilia), microfilaments of actin and myosin (as occurs in muscle cells), and intermediate filaments (each with a distinct protein such as keratin)
Cytosol	The fluid portion of the cytoplasm exclusive of the formed elements listed above; also called hyaloplasm; contains water, minerals, ions, sugars, amino acids, and other nutrients for building macromolecular biopolymers (nucleic acids, proteins, lipids, and large carbohydrates such as starch and cellulose)

Chromosomes

Chromosome Number

In higher organisms, each **somatic** cell (any body cell exclusive of sex cells) contains one set of chromosomes inherited from the **maternal** (female) parent and a comparable set of chromosomes (homologous chromosomes or **homologues**) from the **paternal** (male) parent. The number of chromosomes in this dual set is called the **diploid** ($2n$) number. The suffix "-ploid" refers to chromosome "sets." The prefix indicates the degree of ploidy. Sex cells, or gametes, which contain half the number of chromosome sets found in somatic cells, are referred to as **haploid** cells (n). A **genome** is a set of chromosomes corresponding to the haploid set of a species. The number of chromosomes in each somatic cell is the same for all members of a given species. For example, human somatic cells contain 46 chromosomes, tobacco has 48, cattle 60, the garden pea 14, the fruit fly 8, etc.

Remember

The diploid number of a species bears no direct relationship to the species position in the phylogenetic scheme of classification.

Chromosome Morphology

The structure of chromosomes becomes most easily visible during certain phases of nuclear division when they are highly coiled. Each chromosome in the genome can usually be distinguished from all others by several criteria, including the relative lengths of the chromosomes, the position of a structure called the **centromere** that divides the chromosome into two arms of varying length, the presence and position of enlarged areas called "knobs" or **chromomeres**, the presence of tiny terminal extensions of chromatin material called "satellites," etc. A chromosome with a median centromere (**metacentric**) will have arms of approximately equal size. A **submetacentric**, or **acrocentric**, chromosome has arms of distinctly unequal size. The shorter arm is called the **p arm** and the longer arm is called the **q arm**. If a chromosome has its centromere at or very near one end of the chromosome, it is called **telocentric**. Each chromosome of the genome (with the exception of sex chromosomes) is numbered consecutively according to length, beginning with the longest chromosome first.

Autosomes vs. Sex Chromosomes

In the males of some species, including humans, sex is associated with a morphologically dissimilar (**heteromorphic**) pair of chromosomes called **sex chromosomes**. Such a chromosome pair is usually labeled X and Y. Genetic factors on the Y chromosome determine maleness. Females have two morphologically identical X chromosomes. The members of any other homologous pairs of chromosomes (homologues) are morphologically indistinguishable, but usually are visibly different from other pairs (nonhomologous chromosomes). All chromosomes exclusive of the sex chromosomes are called **autosomes**. Figure 1-2 shows the chromosomal complement of the fruit fly *Drosophila melanogaster* ($2n = 8$) with three pairs of autosomes (2, 3, 4) and one pair of sex chromosomes.

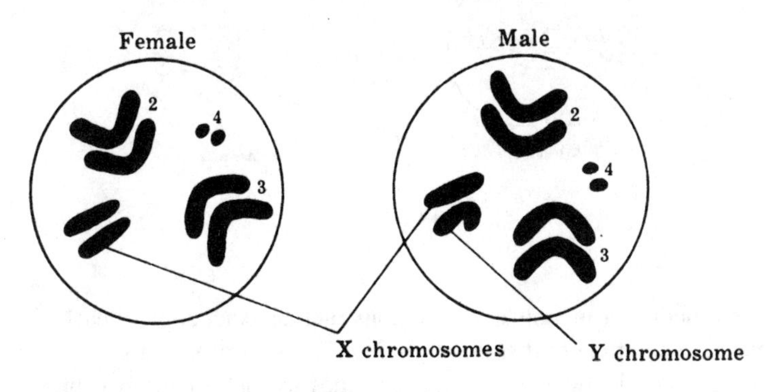

Figure1-2

Cell Division

Mitosis

All somatic cells in a multicellular organism are descendants of one original cell, the fertilized egg, or zygote, through a divisional process called **mitosis** (Figure 1-3).

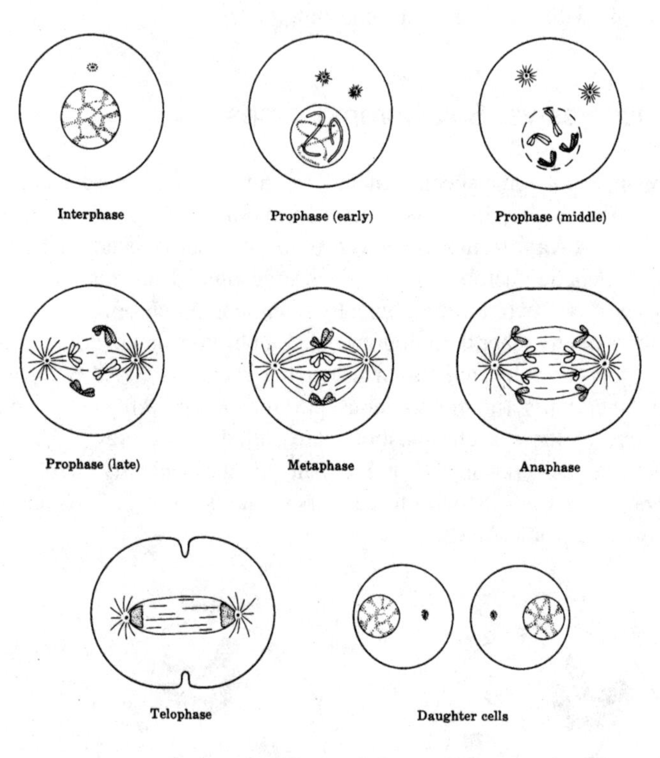

Figure 1-3

The function of mitosis is first to construct an exact copy of each chromosome and then to distribute, through division of the original (mother) cell, an identical set of chromosomes to each of the two progeny cells, or daughter cells. **Interphase** is the period between successive mitoses (Figure 1-4).

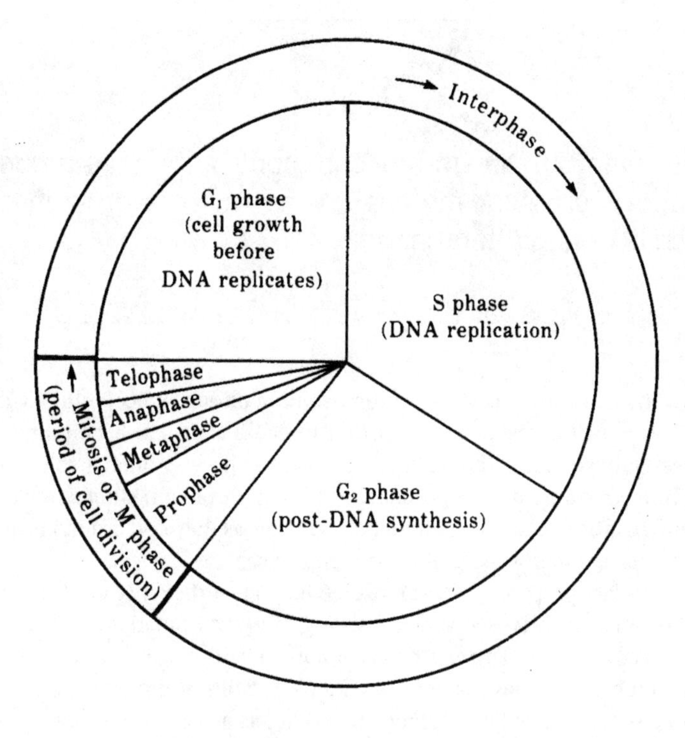

Figure1-4

The double-helix DNA molecule of each chromosome **replicates** during the S phase of the cell cycle (Figure 1-4), producing an identical pair of DNA molecules. Each replicated chromosome thus enters mitosis containing two identical DNA molecules called **chromatids** (sometimes called "sister" chromatids). When DNA associates with histone proteins, it becomes **chromatin** (so called because the complex is readily stained by certain dyes).

A mitotic division has four major phases: **prophase, metaphase, anaphase,** and **telophase**. Within a chromosome, the centromeric regions of each chromatid remain closely associated through the first two phases of mitosis by an unknown mechanism (perhaps by specific centromeric-binding proteins).

Note!

Thin chromatin strands commonly appear as amorphous granular material in the nucleus of stained cells during interphase.

Prophase. In **prophase**, the chromosomes **condense**, becoming visible in the light microscope first as thin threads, and then becoming progressively shorter and thicker. Chromosomes first become visible in the light microscope during prophase. The thin chromatin strands undergo **condensation,** becoming shorter and thicker as they coil around histone proteins and then supercoil upon themselves.

By late prophase, a chromosome may be sufficiently condensed to be seen in the microscope as consisting of two chromatids connected at their centromeres. The **centrioles** of animal cells consist of cylinders of microtubule bundles made of two kinds of **tubulin** proteins. Each centriole is capable of "nucleating" or serving as a site for the construction (mechanism unknown) of a duplicate copy at right angles to itself (Figure 1-1). During prophase, each pair of replicated centrioles migrates toward opposite **polar regions** of the cell and establishes a **microtubule organizing center** (MTOC) from which a spindle-shaped network of microtubules (called the **spindle**) develops. Two kinds of spindle fibers are recognized. Kinetochore microtubules extend from a MTOC to a kinetochore. A **kinetochore** is a fibrous, multiprotein structure attached to centromeric DNA. Polar microtubules extend from a MTOC to some distance beyond the middle of the cell, overlapping in this middle region with similar fibers from the opposite MTOC. Most plants are able to form MTOCs even though they have no centrioles. By late prophase, the nuclear membrane has disappeared and the spindle has fully formed. Late prophase is a good time to study chromosomes (e.g., enumeration) because they are highly condensed and not confined within a nuclear membrane. Mitosis can be arrested at this stage by exposing cells to the alkaloid chemical **colchicine** that interferes with assembly of the spindle fibers. Such treated cells cannot proceed to metaphase until the colchicine is removed.

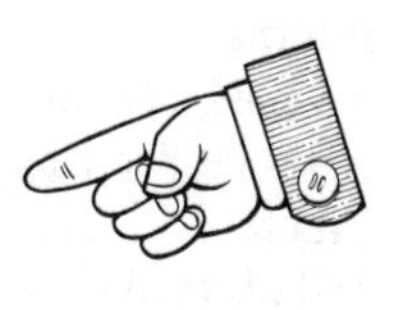

Metaphase. It is hypothesized that during metaphase a dynamic equilibrium is reached by kinetochore fibers from different MTOCs tugging in different directions on the joined centromeres of sister chromatids. This process causes each chromosome to move to a plane near the center of the cell, a position designated the **equatorial plane** or **metaphase plate**. Near the end of metaphase, the concentration of calcium ions increases in the cytosol. Perhaps this is the signal that causes the centromeres of the sister chromatids to dissociate. The exact process remains unknown, but it is commonly spoken of as "division" or "splitting" of the centromeric region.

Anaphase. Anaphase is characterized by the separation of chromatids. According to one theory, the kinetochore microtubules shorten by progressive loss of tubulin subunits, thereby causing former sister chromatids (now recognized as individual chromosomes because they are no longer connected at their centromeres) to migrate toward opposite poles. According to the **sliding filament hypothesis**, with the help of proteins such as **dynein** and **kinesin**, the kinetochore fibers slide past the polar fibers using a ratchet mechanism analogous to the action of the proteins actin and myosin in contracting muscle cells. As each chromosome moves through the viscous cytosol, its arms drag along behind its centromere, giving it a characteristic shape depending upon the location of the centromere.

You Need to Know ✓

Metacentric chromosomes appear V-shaped, submetacentric chromsomes appear J-shaped, and telocentric chromosomes appear rod-shaped.

Telophase. In **telophase**, an identical set of chromosomes is assembled at each pole of the cell. The chromosomes begin to uncoil and return to an interphase condition. The spindle degenerates, the nuclear membrane reforms, and the cytoplasm divides in a process called **cytokinesis**. In animals, cytokinesis is accomplished by the formation of a cleavage furrow that deepens and eventually "pinches" the cell in two as shown in Figure 1-3. Cytokinesis in most plants involves the construction of a **cell plate** of pectin originating in the center of the cell and spreading laterally to the cell wall.

Later, cellulose and other strengthening materials are added to the cell plate, converting it into a new cell wall. The two products of mitosis are called **daughter cells** or **progeny cells** and may or may not be of equal size depending upon where the plane of cytokinesis sections the cell. Thus, while there is no assurance of equal distribution of cytoplasmic components to daughter cells, they do contain exactly the same type and number of chromosomes and hence possess exactly the same genetic constitution.

The time during which the cell is undergoing mitosis is designated the **M period**. The times spent in each phase of mitosis are quite different. Prophase usually requires far longer than the other phases; metaphase is the shortest. DNA replication occurs before mitosis in what is termed the **S** (synthesis) **phase** (Figure 1-4). In nucleated cells, DNA synthesis starts at several positions on each chromosome, thereby reducing the time required to replicate the sister chromatids. The period between M and S is designated the **G_2 phase** (post-DNA synthesis). A long **G_1 phase** (pre-DNA synthesis) follows mitosis and precedes chromosomal replication. Interphase includes G_1, S, and G_2. The four phases (M, G_1, S, and G_2) constitute the life cycle of a somatic cell. The lengths of these phases vary considerably from one cell type to another.

Normal mammalian cells growing in tissue culture usually requires 18–24 hours at 37°C to complete the cell cycle.

Meiosis

Sexual reproduction involves the manufacture of gametes (**gametogenesis**) and the union of a male and a female gamete (**fertilization**) to produce a zygote. Male gametes are **sperms** and female gametes are **eggs**, or **ova** (ovum, singular). Gametogenesis occurs only in the specialized cells (**germ line**) of the reproductive organs (**gonads**). In animals, the testes are male gonads and the ovaries are female gonads. Gametes contain the haploid number (n) of chromosomes, but originate from diploid ($2n$) cells of the germ line. The number of chromosomes must be reduced by half during gametogenesis in order to maintain the chromosome number characteristic of the species. This is accomplished by the divisional process called **meiosis** (Figure 1-5). Meiosis involves a single DNA replication and two divisions of the cytoplasm. The first meiotic division (meiosis I) is a **reductional division** that produces two haploid cells from a single diploid cell. The second meiotic division (meiosis II) is an **equational division** (mitosislike, in that sister chromatids of the haploid cells are separated). Each of the two meiotic divisions consists of four major phases (prophase, metaphase, anaphase, and telophase).

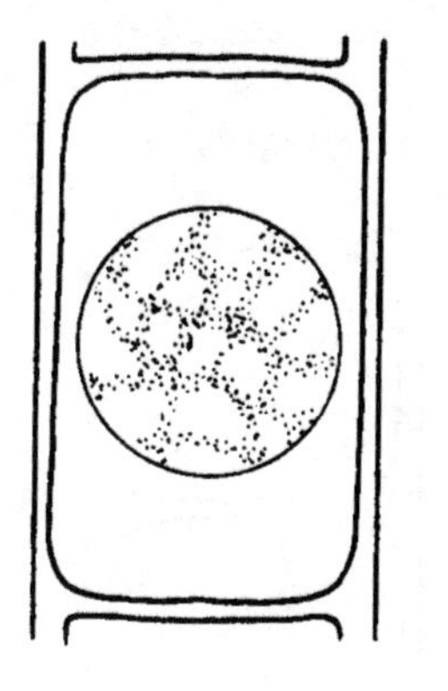

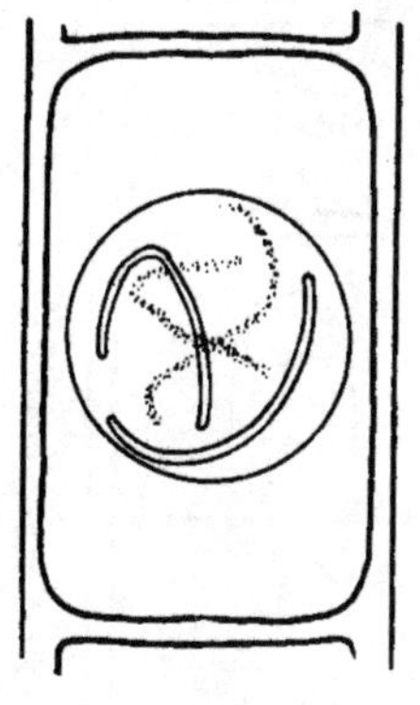

Interphase **Early Prophase I**

Figure 1-5

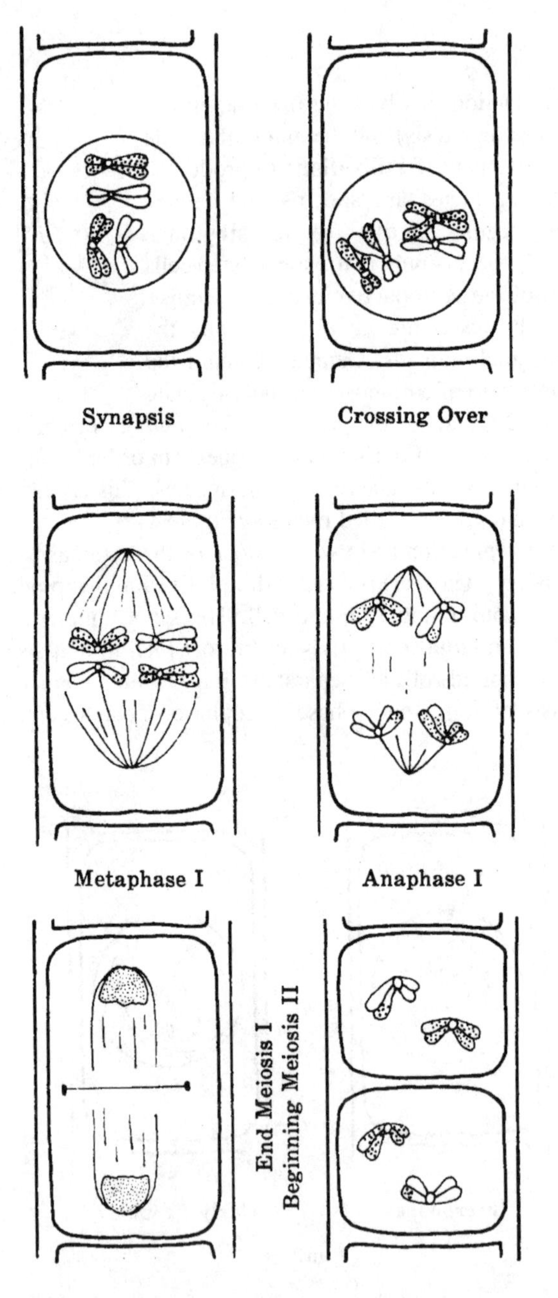
Synapsis
Crossing Over
Metaphase I
Anaphase I
End Meiosis I
Beginning Meiosis II
Telophase I
Prophase II

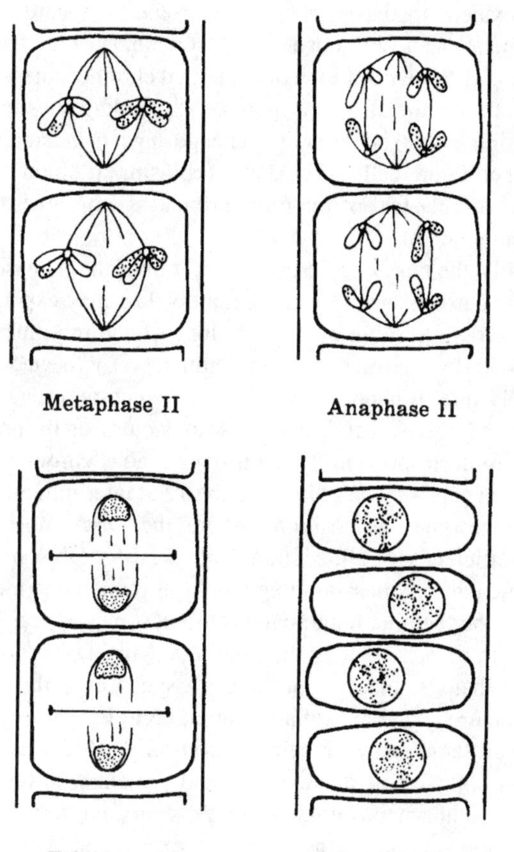

Meiosis I. The DNA replicates during the interphase preceding meiosis It does not replicate between telophase I and prophase II. The prophase of meiosis I differs from the prophase of mitosis in that homologous chromosomes come to lie side by side in a pairing process called **synapsis**. Each pair of synapsed chromosomes is called a **bivalent** (2 chromosomes). Each chromosome consists of two identical sister chromatids at this stage. Thus, a bivalent may also be called a **tetrad** (4 chromatids) if chromatids are counted. The number of chromosomes is always equivalent to the number of centromeres, regardless of how many chromatids each chromosome may contain. During synapsis. nonsister chromatids (one from each of the paired chromosomes) of a tetrad may break and reunite at one or more corresponding sites in a process called **crossing over**. The point of exchange appears in a microscope as a cross-shaped figure called a **chiasma** (**chiasmata**, plural). Thus, at a given chiasma, only two of the four chromatids cross over in a somewhat random manner.

Generally, the number of crossovers per bivalent increases with the length of the chromosome. By chance, a bivalent may experience 0, 1, or multiple crossovers, but even in the longest chromosomes, the incidence of multiple chiasmata of higher numbers is expected to become progressively rare. It is not known whether synapsis occurs by pairing between strands of two different DNA molecules or by proteins that complex with corresponding sites on homologous chromosomes. It is thought that synapsis occurs discontinuously or intermittently along the paired chromosomes at positions where the DNA molecules have unwound sufficiently to allow strands of nonsister DNA molecules to form specific pairs of their building blocks or monomers (nucleotides).

Despite the fact that homologous chromosomes appear in the light microscope to be paired along their entire lengths during prophase I, it is estimated that less than 1% of the DNA synapses in this way. A ribbonlike structure called the **synaptonemal complex** can be seen in the electron microscope between paired chromsomes. It consists of **nucleoprotein** (a complex of nucleic acid and proteins). A few cases are known in which synaptonemal complexes are not formed, but then synapsis is not as complete and crossing over is markedly reduced or eliminated. By the breakage and reunion of nonsister chromatids within a chiasma, linked genes become recombined into **crossover-type** chromatids. The two chromatids within that same chiasma that did not exchange segments maintain the original linkage arrangement of genes as **noncrossover-** or **parental-type** chromatids. A chiasma is a cytological structure visible in the light microscope.

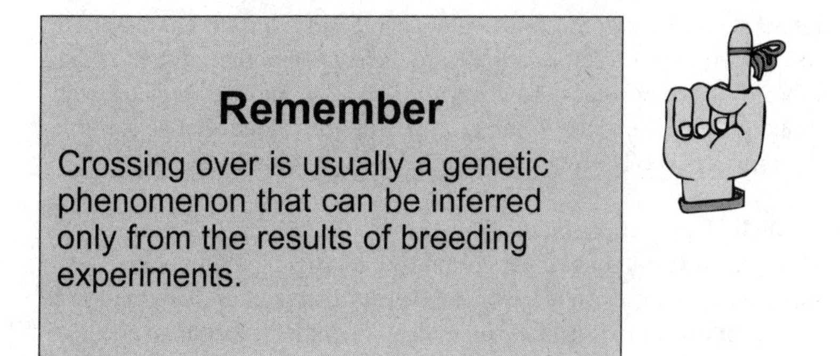

Prophase of meiosis I may be divided into five stages. During **leptonema** (thin-thread stage), the long, thin, attenuated chromosomes start to condense and, as a consequence, the first signs of threadlike structures begin to appear in the formerly amorphous nuclear chromatin material. During **zygonema** (joined-thread stage), synapsis begins. In **pachynema** (thick-thread stage), synapsis appears so tight that it becomes difficult to distinguish homologues in a bivalent. This tight pairing becomes somewhat relaxed during the next stage called **diplonema** (double-thread stage), so that individual chromatids and chiasmata can be seen. Finally, in **diakinesis**, the chromosomes reach their maximal condensation, nucleoli and the nuclear membrane disappear, and the spindle apparatus begins to form.

During **metaphase I**, the bivalents orient at random on the equatorial plane. At **anaphase I**, the centromeres do not divide, but continue to hold sister chromatids together. Because of crossovers, sister chromatids may no longer be genetically identical. Homologous chromosomes separate and move to opposite poles; i.e., whole chromosomes (each consisting of 2 sister chromatids) move apart. This is the movement that will reduce the chromosome number from the diploid ($2n$) condition to the haploid (n) state. Cytokinesis in **telophase I** divides the diploid mother cell into 2 haploid daughter cells. This ends the first meiotic division.

Interkinesis. The period between the first and second meiotic divisions is called **interkinesis**. Depending on the species, interkinesis may be brief or continue for an extended period of time. During an extensive interkinesis, the chromosomes may uncoil and return to an interphase-like condition with reformation of a nuclear membrane. At some later time, the chromosomes would again condense and the nuclear membrane would disappear. Nothing of genetic importance happens during interkinesis. The DNA does *not* replicate during interkinesis!

Meiosis II. In **prophase II**, the spindle apparatus reforms. By **metaphase II**, the individual chromosomes have lined up on the equatorial plane. During **anaphase II**, the centromeres of each chromosome divide, allowing the sister chromatids to be pulled apart in an equatorial division (mitosislike) by the spindle fibers. Cytokinesis in **telophase II** divides each cell into 2 progeny cells. Thus, a diploid mother cell becomes 4 haploid progeny cells as a consequence of a meiotic cycle (meiosis I and meiosis II). The characteristics that distinguish mitosis from meiosis are summarized in Table 1.2.

Mitosis
1. An equational division that separates sister chromatids 2. One division per cycle, i.e., one cytoplasmic division (cytokinesis) per equational chromosomal division 3. Chromosomes fail to synapse; no chiasmata form; genetic exchange between homologous chromosomes does not occur 4. Two products (daughter cells) produced per cycle 5. Genetic content of mitotic products are identical 6. Chromosome number of daughter cells is the same as that of the mother cell 7. Mitotic products are usually capable of undergoing additional mitotic divisions 8. Normally occurs in most all somatic cells 9. Begins at the zygote state and continues through the life of the organism

Meiosis
1. The first stage is a reductional division which separates homologous chromosomes at first anaphase; sister chromatids separate in an equational division at second anaphase
2. Two divisions per cycle, i.e., two cytoplasmic divisions, one following reductional chromosomal division and one following equational chromosomal division
3. Chromosomes synapse and form chiasmata; genetic exchange occurs between homologues
4. Four cellular products (gametes or spores) produced per cycle
5. Genetic content of meiotic products are different; centromeres may be replicas of either maternal or paternal centromeres in varying combinations
6. Chromosome number of meiotic products is half that of the mother cell
7. Meiotic products cannot undergo another meiotic division although they may undergo mitotic division
8. Occurs only in specialized cells of the germ line
9. Occurs only after a higher organism has begun to mature; occurs in the zygote of many algae and fungi

Table 1.2

Mendel's Laws

Gregor Mendel published the results of his genetic studies on the garden pea in 1866 and thereby laid the foundation of modern genetics. In this paper, Mendel proposed some basic genetic principles. One of these is known as the **principle of segregation**. He found that from any one parent, only one allelic form of a gene is transmitted through a gamete to the offspring. For example, a plant which had a factor (or gene) for round-shaped seed and also an allele for wrinkled-shaped seed would transmit only one of these two alleles through a gamete to its offspring. Mendel knew nothing of chromosomes or meiosis, as they had not yet been discovered. We now know that the physical basis for this principle is in first meiotic anaphase where homologous chromosomes segregate or separate from each other. If the gene for round seed is on one chromosome and its allelic form for wrinkled seed is on the homologous chromosome, then it becomes clear that alleles normally will not be found in the same gamete.

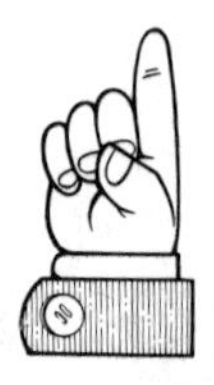

Mendel's **principle of independent assortment** states that the segregation of one factor pair occurs independently of any other factor pair. We know that this is true only for loci on nonhomologous chromosomes. For example, on one homologous pair of chromosomes are the seed shape alleles and on another pair of homologues are the alleles for green and yellow seed color. The segregation of the seed shape alleles occurs independently of the segregation of the seed color alleles because each pair of homologues behaves as an independent unit during meiosis. Furthermore, because the orientation of bivalents on the first meiotic metaphase plate is completely at random, four combinations of factors could be found in the meiotic products: (1) round-yellow, (2) wrinkled-green. (3) round-green, (4) wrinkled-yellow.

Gametogenesis

Usually the immediate end products of meiosis are not fully developed gametes or spores. A period of **maturation** commonly follows meiosis. In plants, one or more meiotic divisions are required to produce reproductive spores, whereas in animals, the meiotic products develop directly into gametes through growth and/or differentiation. The entire process of producing mature gametes or spores, of which meiotic division is the most important, is called **gametogenesis**.

Animal Gametogenesis (*as represented in mammals*)

Gametogenesis in the male animal is called **spermatogenesis** [Figure 1-6(a)]. Mammalian spermatogenesis originates in the germinal epithelium of the seminiferous tubules of the male gonads (testes) from diploid primordial cells. These cells undergo repeated mitotic divisions to form a population of **spermatogonia**. By growth, a spermatogonium may differentiate into a diploid **primary spermatocyte** with the capacity to undergo meiosis. The first meiotic division occurs in these primary spermatocytes, producing haploid **secondary spermatocytes**. From these cells, the second meiotic division produces 4 haploid meiotic products called **spermatids**. Almost the entire amount of cytoplasm then extrudes into a long whiplike tail during maturation and the cell becomes transformed into a mature male gamete called a **sperm cell** or **spermatazoan** (-zoa, plural).

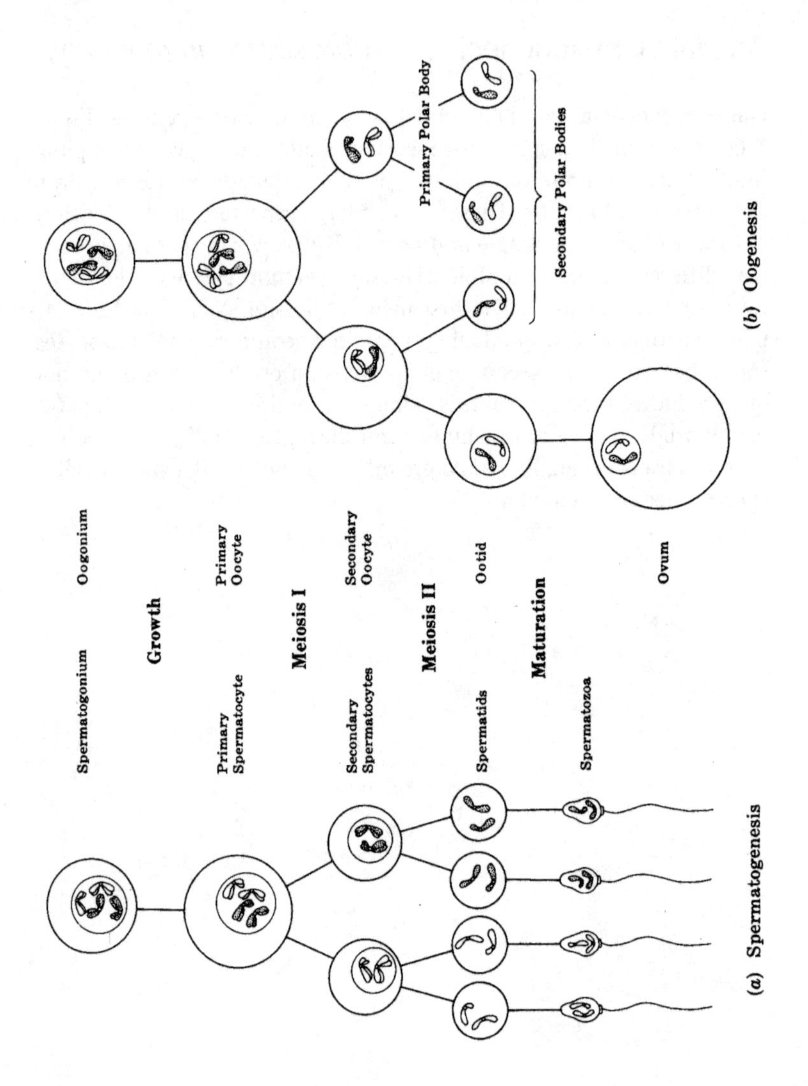
Spermatogonium
Growth
Oogonium
Primary Spermatocyte
Meiosis I
Primary Oocyte
Secondary Spermatocytes
Meiosis II
Secondary Oocyte
Spermatids
Maturation
Ootid
Spermatozoa
Ovum
Primary Polar Body
Secondary Polar Bodies
(a) Spermatogenesis
(b) Oogenesis

Figure 1-6

Gametogenesis in the female is called **oogenesis** [Figure 1-6(b)]. Mammalian oogenesis originates in the germinal epithelium of the female gonads (ovaries) in diploid primordial cells called **oogonia**. By growth and storage of much cytoplasm or yolk (to be used by the early embryo), the oogonium is transformed into a diploid **primary oocyte** with the capacity to undergo meiosis. The first meiotic division reduces the chromosome number by half and also distributes vastly different amounts of cytoplasm to the two products by a grossly unequal cytokinesis. The larger cell thus produced is called a **secondary oocyte** and the smaller is a primary **polar body**. In some cases, the first polar body may undergo the second meiotic division, producing two secondary polar bodies. All polar bodies degenerate, however, and take no part in fertilization. The second meiotic division of the oocyte again involves an unequal cytokinesis, producing a large, yolky **ootid** and a secondary polar body. By additional growth and differrentiation, the ootid becomes a mature female gamete called an **ovum** or **egg cell**.

The union of male and female gametes (sperm and egg) is called fertilization and reestablishes the diploid number in the resulting cell called a zygote. The head of the sperm enters the egg, but the tail piece (the bulk of the cytoplasm of the male gamete) remains outside and degenerates. Subsequent mitotic divisions produce the numerous cells of the embryo that become organized into the tissues and organs of the new individual.

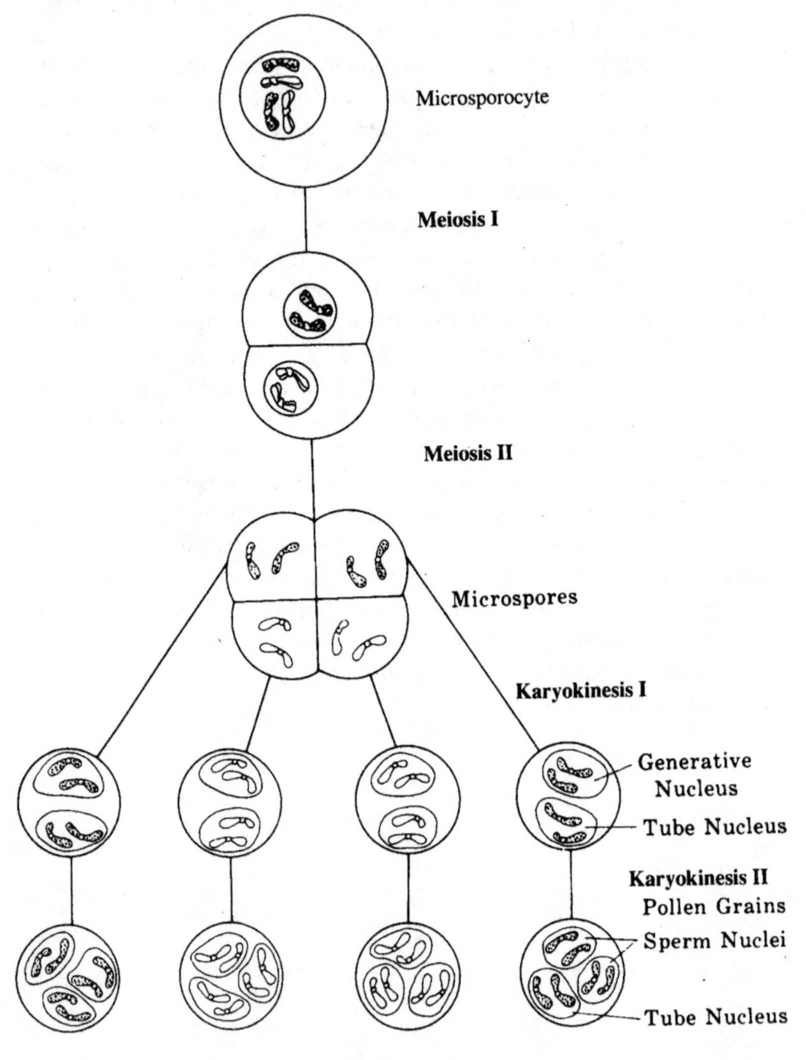
Microsporocyte
Meiosis I
Meiosis II
Microspores
Karyokinesis I
Generative Nucleus
Tube Nucleus
Karyokinesis II
Pollen Grains
Sperm Nuclei
Tube Nucleus

Figure 1-7

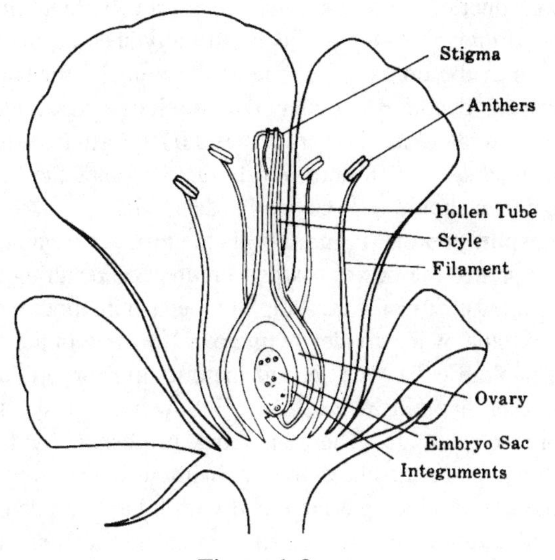

Figure 1-8

Plant Gametogenesis (*as represented in angiosperms*)

Gametogenesis in the plant kingdom varies considerably between major groups of plants. The process as described below is that typical of many flowering plants (**angiosperms**). **Microsporogenesis** (Figure 1-7) is the process of gametogenesis in the male part of the flower (**anther**, Figure1-8) resulting in reproductive spores called **pollen grains**. A diploid microspore mother cell (**microsporocyte**) in the anther divides by meiosis, forming at the first division a pair of haploid cells. The second meiotic division produces a cluster of four haploid **microspores**. Following meiosis, each microspore undergoes a mitotic division of the chromosomes without a cytoplasmic division (**karyokinesis**). This requires chromosomal replication that is not illustrated in the karyokinetic divisions of Figure

1-7. The product of the first karyokinesis is a cell containing two identical haploid nuclei. Pollen grains are usually shed at this stage. Upon germination of the pollen tube, one of these nuclei (or haploid sets of chromosomes) becomes a **generative nucleus** and divides again by mitosis without cytokinesis (karyokinesis II) to form two **sperm nuclei**. The other nucleus, which does not divide, becomes the **tube nucleus**. All three nuclei should be genetically identical.

Megasporogenesis (Figure 1-9) is the process of gametogenesis in the female part of the flower (**ovary**, Figure 1-8) resulting in reproductive cells called **embryo sacs**. A diploid megaspore mother cell (**megasporocyte**) in the ovary divides by meiosis, forming in the first division a pair of haploid cells. The second meiotic division produces a linear group of four haploid *megaspores*. Following meiosis, three of the megaspores degenerate. The remaining megaspore undergoes three mitotic divisions of the chromosomes without intervening cytokineses (karyokineses), producing a large cell with 8 haploid nuclei (immature embryo sac). Remember that chromosomal replication must precede each karyokinesis, but this is not illustrated in Figure1-9. The sac is surrounded by maternal tissues of the ovary called **integuments** and by the megaspoangium (**nucellus**). At one end of the sac, there is an opening in the integuments (**micropyle**) through which the pollen tube will penetrate. Three nuclei of the sac orient themselves near the micropylar end and two of the three (**synergids**) degenerate. The third nucleus develops into an **egg nucleus**. Another group of three nuclei moves to the opposite end of the sac and degenerates (**antipodals**). The two remaining nuclei (**polar nuclei**) unite near the center of the sac, forming a single diploid **fusion nucleus**. The mature embryo sac (**megagametophyte**) is now ready for fertilization.

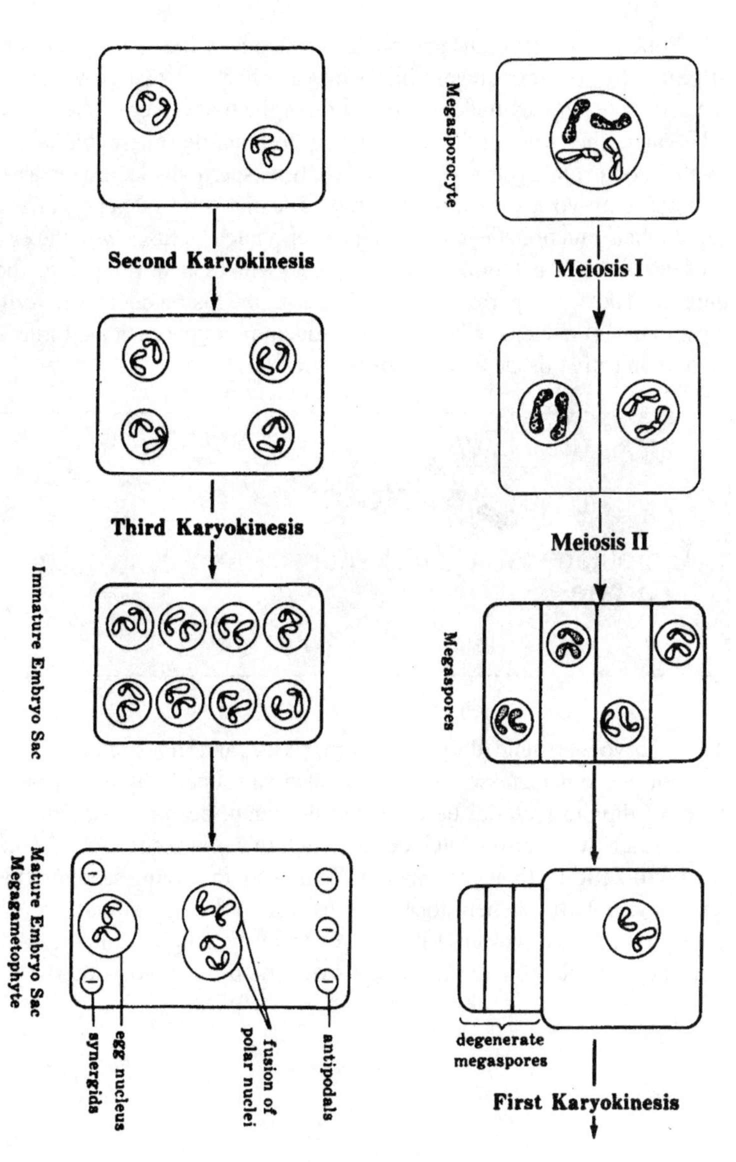
Megasporocyte
Meiosis I
Meiosis II
Megaspores
degenerate
megaspores
First Karyokinesis
Second Karyokinesis
Third Karyokinesis
Immature Embryo Sac
Mature Embryo Sac
Megagametophyte
synergids
egg nucleus
fusion of
polar nuclei
antipodals

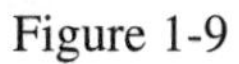
Figure 1-9

Pollen grains from the anthers are carried by wind or insects to the **stigma**. The pollen grain germinates into a pollen tube that grows down the **style**, presumably under the direction of the tube nuceus. The pollen tube enters the ovary and makes its way through the micropyle of the ovule into the embryo sac (Figure 1-10). Both sperm nuclei are released into the embryo sac. The pollen tube and the tube nucleus, having served their function, degenerate. One sperm nucleus fuses with the egg nucleus to form a diploid zygote, which will then develop into the embryo. The other sperm nucleus unites with the fusion nucleus to form a triploid ($3n$) nucleus, which, by subsequent mitotic divisions, forms a starchy nutritive tissue called **endosperm**.

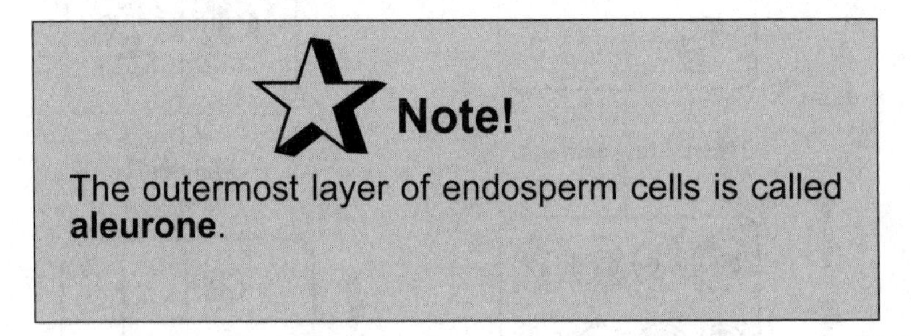

The embryo, surrounded by endosperm tissue, and in some cases such as corn and other grasses where it is also surrounded by a thin outer layer of diploid material tissue called **pericarp**, becomes the familiar seed. Since two sperm nuclei are involved, this process is termed **double fertilization**. Upon germination of the seed, the young seedling (the next sporophytic generation) utilizes the nutrients stored in the endosperm for growth until it emerges from the soil, at which time it becomes capable of manufacturing its own food by photosynthesis.

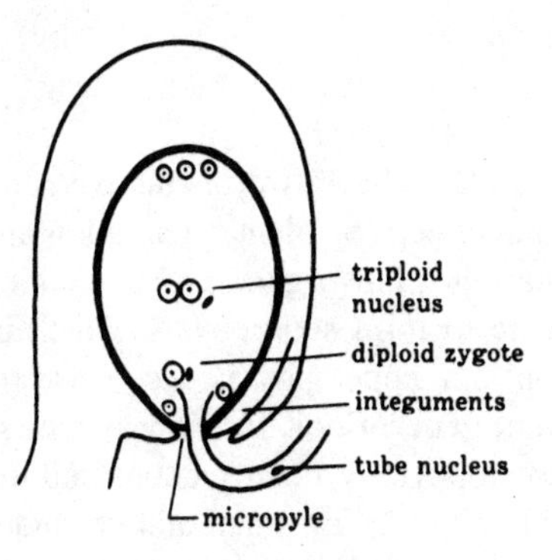
triploid
nucleus
diploid zygote
integuments
tube nucleus
micropyle
Fertilization

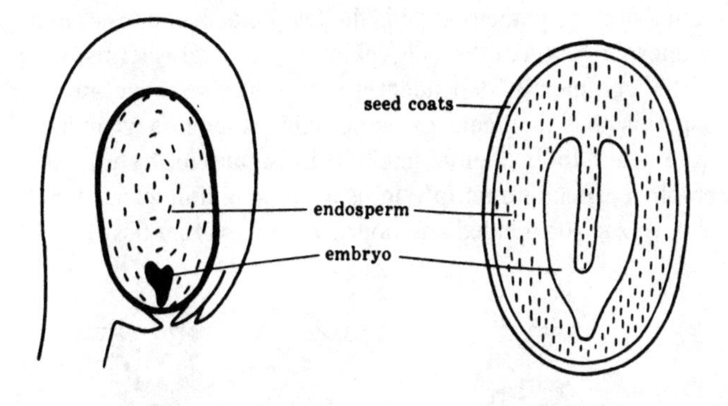
seed coats
endosperm
embryo
Developing Seed
Mature Seed

Figure 1-10

Life Cycles

Life cycles of most plants have two distinctive generations: a haploid **gametophytic** (gamete-bearing plant) generation and a diploid **sporophytic** (spore-bearing plant) generation. Gametophytes produce gametes which unite to form sporophytes, which in turn give rise to spores that develop into gametophytes, etc. This process is referred to as the **alternation of generations**. In lower plants, such as mosses and liverworts, the gametophyte is a conspicuous and independently living generation, the sporophyte being small and dependent upon the gametophyte. In higher plants (ferns, gymnosperms, and angiosperms), the situation is reversed; the sporophyte is the independent and conspicuous generation and the gametophyte is the less conspicuous and, in the case of gymnosperms (cone-bearing plants) and angiosperms (flowering plants), a completely dependent generation. We have just seen in angiosperms that the male gametophytic generation is reduced to a pollen tube and three haploid nuclei (**microgametophyte**). The female gametophyte (**megagametophyte**) is a single multinucleated cell called the embryo sac surrounded and nourished by ovarian tissue.

You Need to Know ✓

Many simpler organisms such as one-celled animals (protozoa), algae, yeast, and other fungi are useful in genetic studies and have interesting life cycles that exhibit considerable variation.

Solved Problem 1.1. Develop a general formula that expresses the number of different types of genetic chromosomal combinations which can be formed in an organism with k pairs of chromosomes.

Solution. It is known that 1 pair of chromsomes gives 2 types of gametes, 2 pairs gives 4 types of gametes, 3 pairs give 8 types, etc. The progression 2, 4, 8, ... can be expressed by the formula 2^k, where k is the number of chromosome pairs.

Solved Problem 1.2. When a plant of chromosomal type *aa* pollinates a plant of type *AA*, what chromsomal type of embryo and endosperm is expected in the resulting seeds?

Solution. The pollen parent produces two sperm nuclei in each pollen grain of type *a*, one combining with *A* egg nucleus to produce a diploid zygote (embryo) of type *Aa* and the other combining with the maternal fusion nucleus *AA* to produce a triploid endosperm of type *AAa*.

Solved Problem 1.3. Given the first meiotic metaphase orientation shown below, and keeping all products in sequential order as they would be formed from left to right, diagram the embryo sac that develops from the meiotic product at the left and label the chromsomal constitution of all its nuclei.

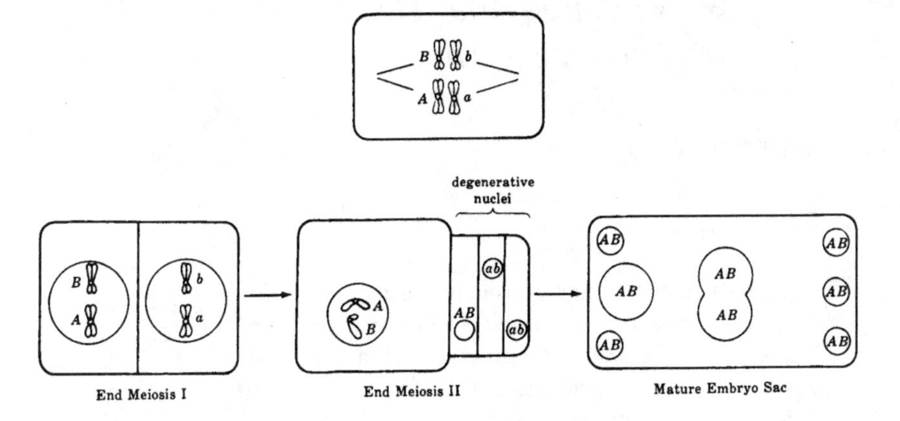

Chapter 2
SINGLE-GENE INHERITANCE

IN THIS CHAPTER:

- ✔ *Terminology*
- ✔ *Allelic Relationships*
- ✔ *Single-Gene (Monofactorial) Crosses*
- ✔ *Pedigree Analysis*
- ✔ *Probability Theory*

Terminology

Phenotype

A **phenotype** may be any measurable characteristic or distinctive trait possessed by an organism. The trait may be visible to the eye, such as the color of a flower or the texture of hair, or it may require special tests for identification, as in the determination of the respiratory quotient or the serological test for blood type. The phenotype is the result of gene products brought to expression in a given environment.

Example 2.1 Rabbits of the Himalayan breed in the usual range of environments develop black pigment at the tips of the nose, tail, feet and ears. If raised at very high temperatures, an all-white rabbit is produced. The gene for Himalayan color pattern specifies a temperature sensitive enzyme that is inactivated at high temperature, resulting in a loss of pigmentation.

The kinds of traits that we shall encounter in the study of simple Mendelian inheritance will be considered to be relatively unaffected by the normal range of environmental conditions in which the organism is found. It is important, however, to remember that genes establish boundaries within which the environment may modify the phenotype.

Genotype

All of the genes possessed by an individual constitute its **genotype**. In this chapter, we shall be concerned only with that portion of the genotype involving alleles at a single locus.

Homozygous. The union of gametes carrying identical alleles produces a **homozygous** genotype. A homozygote produces only one kind of gamete.

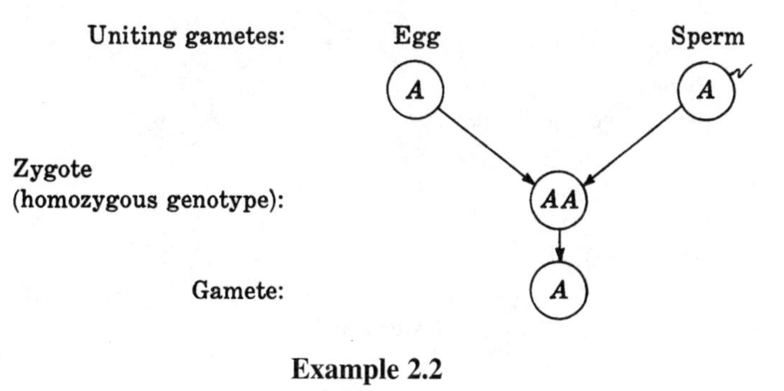

Example 2.2

Pure Line. A group of individuals with similar genetic background (breeding) is often referred to as a line or strain or variety or breed. Self-fertilization or mating closely related individuals for many generations (inbreeding) usually produces a population that is homozygous at nearly all loci. Matings between the homozygous individuals of a **pure line** produce only homozygous offspring like the parents. Thus, we say that a pure line "breeds true."

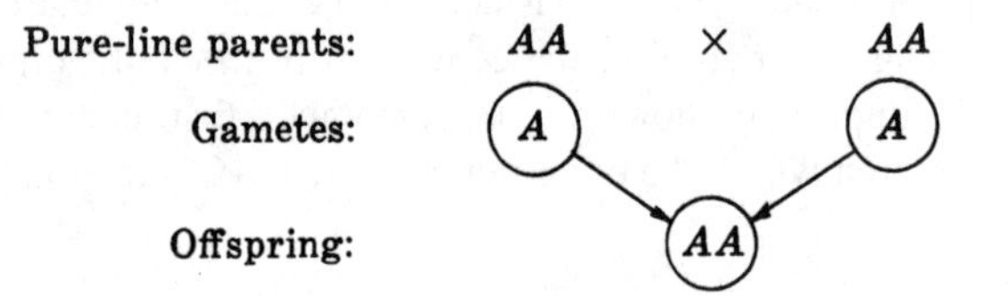

Example 2.3

Heterozygous. The union of gametes carrying different alleles produces a **heterozygous** genotype. Different kinds of gametes are produced by a heterozygote.

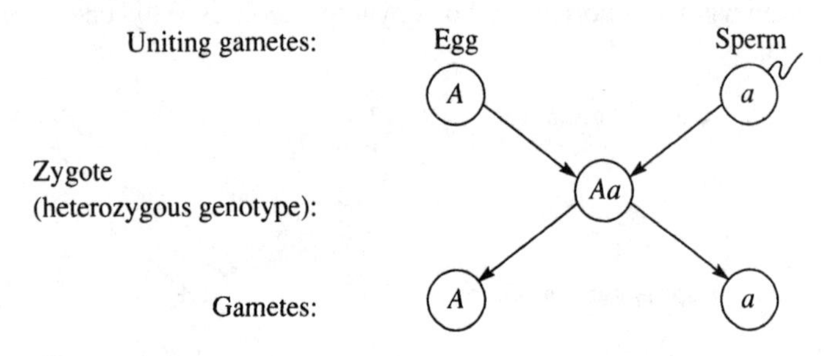

Example 2.4

Hybrid. The term hybrid is synonymous with the heterozygous condition. This chapter may involve a single-factor hybrid (monohybrid) while the next chapter will consider heterozygosity at two or more loci (polyhybrids).

Allelic Relationships

Dominant and Recessive Alleles

Whenever one of a pair of alleles can come to phenotype expression only in a homozygous genotype, we call that allele a **recessive** factor. The allele that can phenotypically express itself in the heterozygote as well as in the homozygote is called a **dominant** factor.

Upper- and lowercase letters are commonly used to designate dominant and recessive alleles, respectively. Usually, the genetic symbol corresponds to the first letter in the name of the abnormal (or mutant) trait.

Example 2.5 Lack of pigment deposition in the human body is an abnormal recessive trait called "albinism." Using *A* and *a* to represent the dominant (normal) allele and the recessive (albino) allele, respectively, three genotypes and two phenotypes are possible:

Genotype	Reaction with:		Blood Group
	Anti-M	Anti-N	(Phenotype)
$L^M L^M$	+	–	M
$L^M L^N$	+	+	MN
$L^N L^N$	–	+	N

Carriers. Recessive alleles (such as the one for albinism) are often deleterious to those who possess them in duplicate (homozygous recessive genotype). A heterozygote may appear just as normal as the homozygous dominant genotype. A heterozygous individual who possesses a deleterious recessive allele hidden from phenotypic expression by the dominant normal allele is called a **carrier**.

Remember

Most of the deleterious alleles harbored by a population are found in carrier individuals.

Wild-Type Symbolism. A different system for symbolizing dominant and recessive alleles is widely used in numerous organisms from higher plants and animals to the bacteria and viruses. Where one phenotype is obviously of much more common occurrence in the population than its alternative phenotype, the former is usually referred to as **wild type**.

The phenotype that is rarely observed is called the **mutant type**. In this system, the symbol + is used to indicate the normal allele for wild type. The base letter for the gene usually is taken from the name of the mutant or abnormal trait. If the mutant gene is recessive, the symbol would be a lowercase letter(s) corresponding to the initial letter(s) in the name of the trait. Its normal (wild-type) dominant allele would have the same lowercase letter but with a + as a superscript.

Example 2.6 Black body color in *Drosophila* is governed by a recessive gene b, and wild-type (gray body) by its dominant allele b^+.

If the mutant trait is dominant, the base symbol would be an uppercase letter without a superscript, and its recessive wild-type allele would have the same uppercase symbol with a + as a superscript.

Example 2.7 Lobe-shaped eyes in *Drosophila* are governed by a dominant gene L, and wild-type (oval eye) by its recessive allele L^+.

Remember that the case of the symbol indicates the dominance or recessiveness of the *mutant* allele to which the superscript + for wild type must be referred. After the allelic relationships have been defined, the symbol + by itself may be used for wild type and the letter alone may designate the mutant type.

Codominant Alleles

Alleles that lack dominant and recessive relationships may be called incompletely dominant, partially dominant, semidominant, or **codominant**. This means that each allele is capable of some degree of expression when in the heterozygous condition. Hence the heterozygous genotype gives rise to a phenotype distinctly different from either of the homozygous genotypes. Usually the heterozygous phenotype resulting from codominance is intermediate in character between those produced by the homozygous genotypes; hence the erroneous concept of "blending." The phenotype may appear to be a "blend" in heterozygotes, but the alleles maintain their individual identities and will segregate from each other in the formation of gametes.

Symbolism for Codominant Alleles. For codominant alleles, all uppercase base symbols with different subscripts should be used. The uppercase letters call attention to the fact that each allele can express itself to some degree even when in the presence of its alternative allele (heterozygous).

Example 2.8 The alleles governing the M-N blood group system in humans are codominants and may be represented by the symbols L^M and L^N, the base letter (L) being assigned in honor of its discoverers (Landsteiner and Levine). Two antisera (anti-M and anti-N) are used to distinguish three genotypes and their corresponding phenotypes (blood groups). Agglutination is represented by + and nonagglutination by –.

Genotype	Reaction with:		Blood Group
	Anti-M	Anti-N	(Phenotype)
L^ML^M	+	–	M
L^ML^N	+	+	MN
L^NL^N	–	+	N

Lethal Alleles

The phenotypic manifestation of some genes is the death of the individual in either the prenatal or postnatal period prior to maturity. Such factors are called **lethal genes**. A fully dominant lethal allele (i.e., one that kills in both the homozygous and heterozygous conditions) occasionally arises by mutation from a normal allele. Individuals with a dominant lethal allele die before they can leave progeny. Therefore, the mutant dominant lethal allele is removed from the population in the same generation in which it arose. Lethals that kill only when homozygous may be of two kinds: (1) one that has no obvious phenotypic effect in heterozygotes, and (2) one that exhibits a distinctive phenotype when heterozygous.

Example 2.9 By special techniques, a completely recessive lethal (*l*) can sometimes be identified in certain families.

Genotype	Phenotype
LL, *Ll*	Normal viability
ll	Lethal

Example 2.10 The amount of chlorophyll in snapdragons (*Antirrhinum*) is controlled by a pair of codominant alleles, one of which exhibits a lethal effect when homozygous, and a distinctive color phenotype when heterozygous.

Genotype	Phenotype
C^1C^1	Green (normal)
C^1C^2	Pale green
C^2C^2	White (lethal)

Penetrance and Expressivity

Differences in environmental conditions or in genetic backgrounds may cause individuals that are genetically identical at a particular locus to exhibit different phenotypes. The percentage of individuals with a particular gene combination that exhibits the corresponding character to any degree represents the **penetrance** of the trait.

Example 2.11 In some families, extra fingers and/or toes (polydactyly) in humans is thought to be produced by a dominant gene (*P*). The normal condition with five digits on each limb is produced by the recessive genotype (*pp*). Some individuals of genotype *Pp* are not polydactylous, and therefore the gene has a penetrance of less than 100 percent.

A trait, although penetrant, may be quite variable in its expression. The degree of effect produced by a penetrant genotype is termed **expressivity**.

Example 2.12 The polydactylous condition may be penetrant in the left hand (six fingers) and not in the right (five fingers), or it may be penetrant in the feet and not in the hands.

A recessive lethal gene that lacks complete penetrance and expressivity will kill less than 100 percent of the homozygotes before sexual maturity. The terms **semilethal** or **subvital** apply to such genes. The effects that various kinds of lethals have on the reproduction of the next generation form a broad spectrum from complete lethality to sterility in completely viable genotypes. This book will consider only those lethals that become completely penetrant, usually during the embryonic stage.

Multiple Alleles

The genetic systems proposed thus far have been limited to a single pair of alleles. The maximum number of alleles at a gene locus that any individual possesses is two, with one on each of the homologous chromosomes. But since a gene can be changed to alternative forms by the process of mutation, a large number of alleles is theoretically possible in a population of individuals. Whenever more than two alleles are identified at a gene locus, we have a **multiple allelic series.**

Symbolism for Multiple Alleles. The dominance hierarchy should be defined at the beginning of each problem involving multiple alleles. A capital letter is commonly used to designate the allele that is dominant to all others in the series. The corresponding lowercase letter designates the allele that is recessive to all others in the series. Other alleles, intermediate in their degree of dominance between these two extremes, are usually assigned the lowercase letter with some suitable superscript.

Example 2.13 A classical example of multiple alleles is found in the ABO blood group system of humans, where the allele I^A for the A antigen is codominant with the allele I^B for the B antigen. Both I^A and I^B are completely dominant to the allele i, which fails to specify any detectable antigenic structure. The hierarchy of dominance relationships is symbolized as $(I^A = I^B) > i$. Two antisera (anti-A and anti-B) are required for the detection of four phenotypes.

Genotypes	Reaction with:		Phenotype
	Anti-A	Anti-B	(Blood Groups)
I^AI^A, I^Ai	+	−	A
I^BI^B, I^Bi	−	+	B
I^AI^B	+	+	AB
ii	−	−	O

Single-Gene (Monofactorial) Crosses

The Six Basic Types of Matings

A pair of alleles governs pelage color in the guinea pig; a dominant allele B produces black and its recessive allele b produces white. There are 6 types of matings possible among the 3 genotypes. The parental generation is symbolized P and the first filial generation of offspring is symbolized $\mathbf{F}_1$.

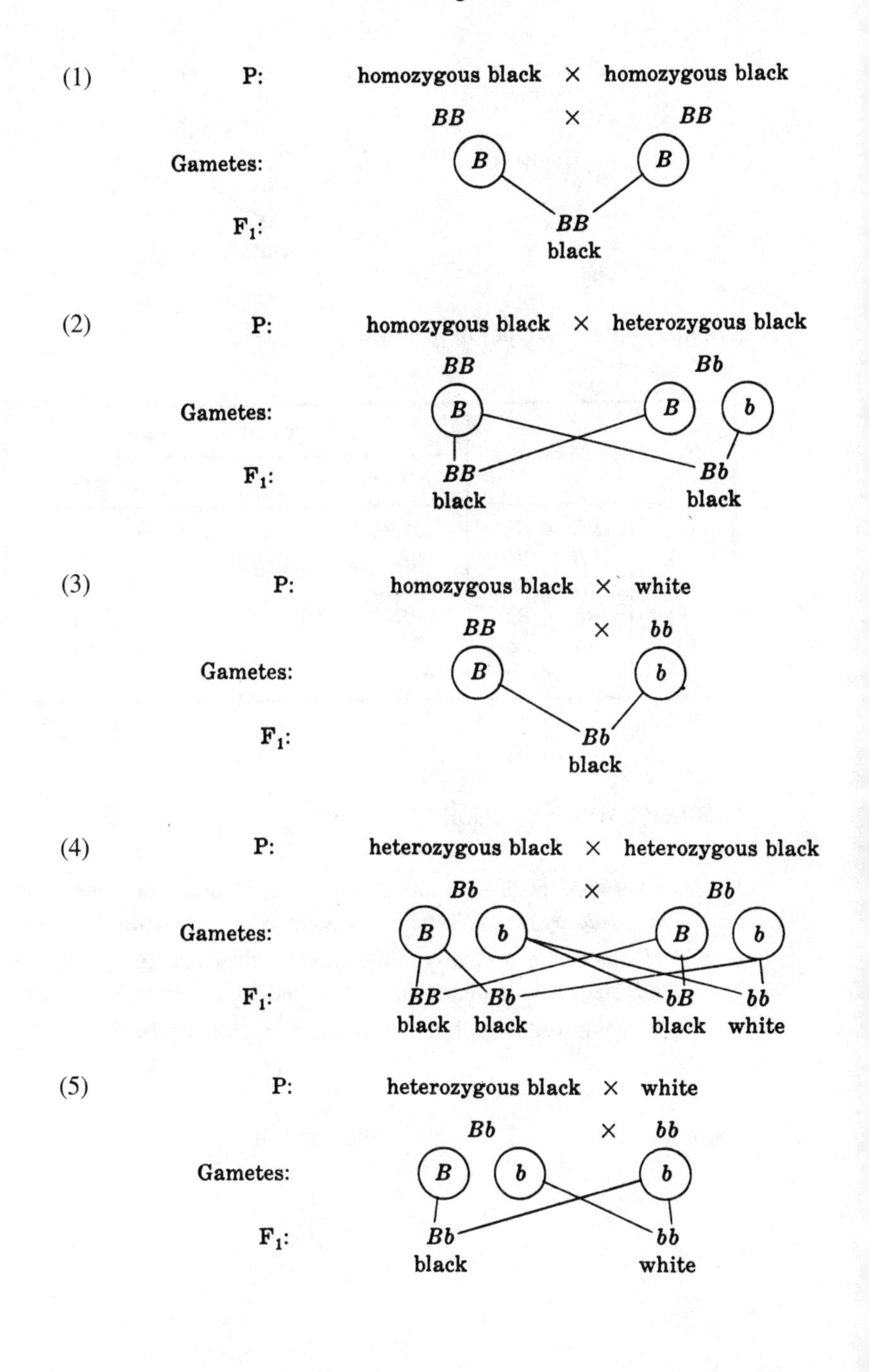
(1)
P: homozygous black × homozygous black
BB × BB
Gametes: B B
F1: BB
black
(2)
P: homozygous black × heterozygous black
BB Bb
Gametes: B B b
F1: BB Bb
black black
(3)
P: homozygous black × white
BB × bb
Gametes: B b
F1: Bb
black
(4)
P: heterozygous black × heterozygous black
Bb × Bb
Gametes: B b B b
F1: BB Bb bB bb
black black black white
(5)
P: heterozygous black × white
Bb × bb
Gametes: B b b
F1: Bb bb
black white

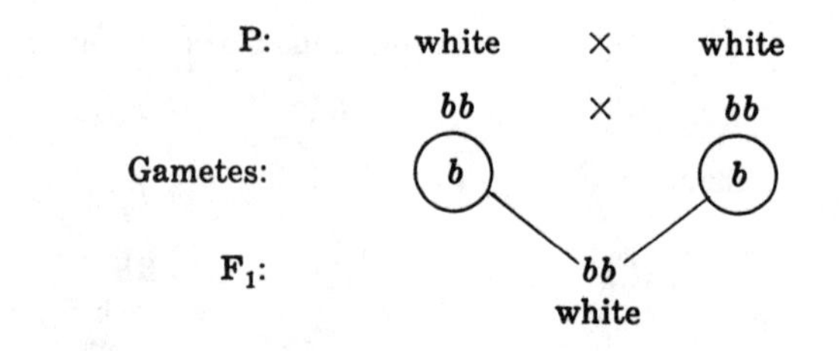

Summary of the six types of matings:

No.	Matings	Expected F_1 Ratios	
		Genotypes	Phenotypes
(1)	$BB \times BB$	All BB	All black
(2)	$BB \times Bb$	$\frac{1}{2}BB : \frac{1}{2}Bb$	All black
(3)	$BB \times bb$	All Bb	All black
(4)	$Bb \times Bb$	$\frac{1}{4}BB : \frac{1}{2}Bb : \frac{1}{4}bb$	$\frac{3}{4}$ black : $\frac{1}{4}$ white
(5)	$Bb \times bb$	$\frac{1}{2}Bb : \frac{1}{2}bb$	$\frac{1}{2}$ black : $\frac{1}{2}$ white
(6)	$bb \times bb$	All bb	All white

Conventional Production of the F_2

Unless otherwise specified in the problem, the second filial generation (**F_2**) is produced by crossing the F_1 individuals among themselves randomly. If plants are normally self-fertilized, they can be artificially cross-pollinated in the parental generation and the resulting F_1 progeny may then be allowed to pollinate themselves to produce the F_2.

Example 2.14 P: $BB \times bb$
black white

F_1: Bb
black

The black F_1 males are mated to the black F_1 females to produce the F_2.

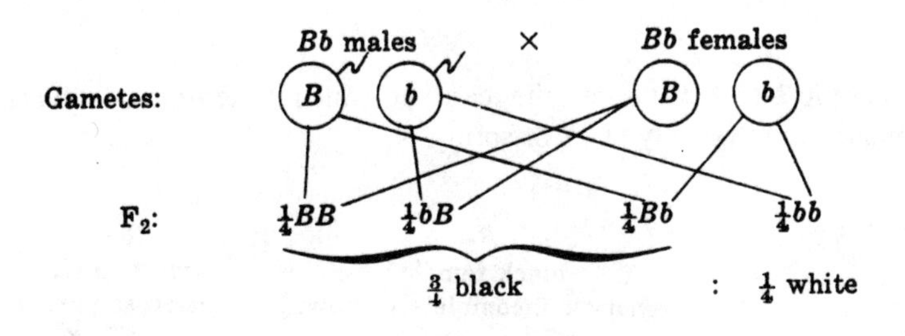

An alternative method for combining the F_1 gametes is to place the female gametes along one side of a "checkerboard" (**Punnett square**) and the male gametes along the other side and then combine them to form zygotes as shown below.

	B	*b*
B	*BB* black	*Bb* black
b	*Bb* black	*bb* white

Testcross

Because a homozygous dominant genotype has the same phenotype as the heterozygous genotype, a **testcross** is required to distinguish between them. The testcross parent is always homozygous recessive for all of the genes under consideration. The purpose of a testcross is to discover how many different kinds of gametes are being produced by the individual whose genotype is in question. A homozygous dominant individual will produce only one kind of gamete; a **monohybrid** individual (heterozygous at one locus) produces two kinds of gametes with equal frequency.

Example 2.15 Consider the case in which testcrossing a black female produced only black offspring.

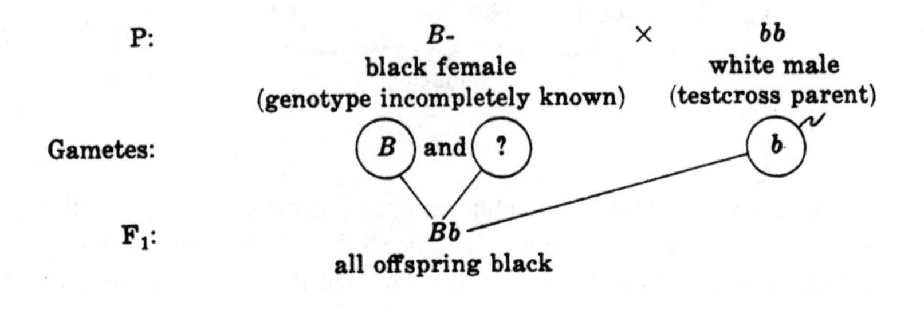

Conclusion: The female parent must be producing only one kind of gamete and therefore she is homozygous dominant *BB*.

Example 2.16 Consider the case in which testcrossing a black male produced black and white offspring in approximately equal numbers.

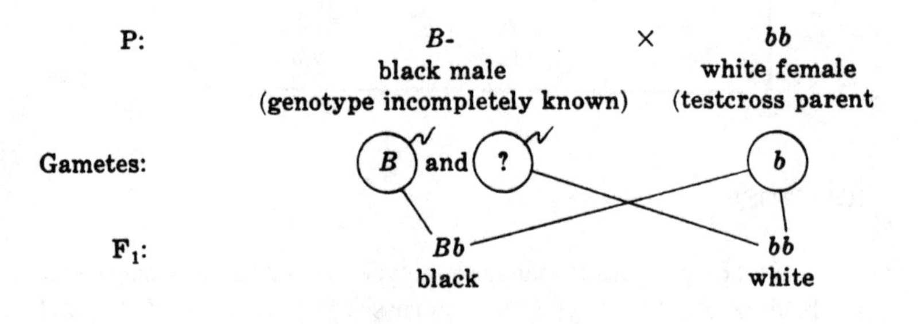

Conclusion: The male parent must be producing 2 kinds of gametes and therefore he is heterozygous *Bb*.

Backcross

If the F_1 progeny is mated back to one of its parents (or to individuals with a genotype identical to that of their parents), the mating is termed **backcross**.

Pedigree Analysis

A pedigree is a systematic listing (either as words or as symbols) of the ancestors of a given individual, or it may be the "family tree" for a large number of individuals. It is customary to represent females as circles and males as squares. Matings are shown as horizontal lines between two individuals. The offspring of a mating are connected by a vertical line to the mating line. Different shades or colors added to the symbols can represent various phenotypes. Each generation is listed on a separate row labeled with Roman numerals. Individuals within a generation receive Arabic numerals.

Example 2.17 Let solid symbols represent black guinea pigs and open symbols represent white guinea pigs.

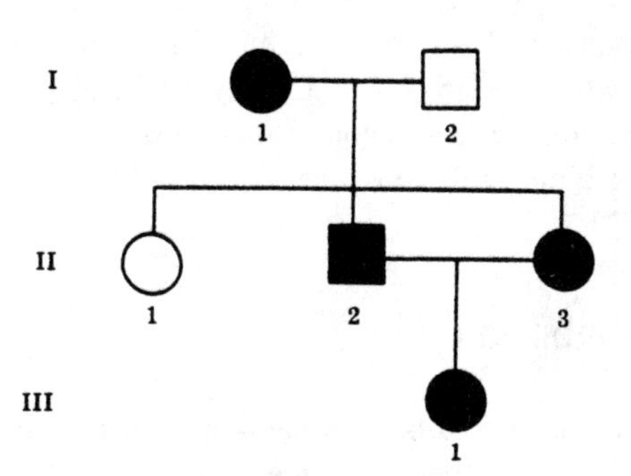

Individuals	Phenotype	Genotype
I1	Black ♀	*Bb*
I2	White ♂	*bb*
II1	White ♀	*bb*
II2	Black ♂	*Bb*
II3	Black ♀	*Bb*
III1	Black ♀	*B*-*

* The dash indicates that the genotype could be either homozygous or heterozygous.

Probability Theory

Observed vs. Expected Results

Experimental results seldom conform exactly to the expected ratios. Genetic probabilities derive from the operation of chance events in the meiotic production of gametes and the random union of these gametes in fertilization. Samples from a population of individuals often deviate from the expected ratios, rather widely in very small samples, but usually approaching the expectations more closely with increasing sample size.

Example 2.18 Suppose that a testcross of heterozygous black guinea pigs ($Bb \times bb$) produces 5 offspring: 3 black (*Bb*) and 2 white (*bb*). Theoretically, we expect half of the total number of offspring to be black and half to be white = ½ (5) = 2 ½. Obviously, we cannot observe half of an individual, and the results conform as closely to the theoretical expectations as is biologically possible.

Combining Probabilities

Two or more events are said to be **independent** if the occurrence or nonoccurrence of any one of them does not affect the probabilities of occurrence of any of the others. When 2 independent events occur with the probabilities p and q, respectively, then the probability of their joint occurrence is pq. That is, the combined probability is the product of the probabilities of the independent events. If the word "and" is used or implied in the phrasing of a problem solution, a *multiplication* of independent probabilities is usually required.

Example 2.19 In testcrossing a heterozygous black guinea pig ($Bb \times bb$), let the probability of a black (Bb) offspring be $p = 1/2$ and of a white (bb) offspring be $q = 1/2$. The combined probability of the first two offspring being white (i.e., the first offspring is white *and* the second offspring is white) $= q \times q = q^2 = (1/2)^2 = 1/4$.

There is only one way in which two heads may appear in two tosses of a coin, i.e., heads on the first toss and heads on the second toss. The same is true for two tails. There are two ways. However, to obtain one head and one tail in two tosses of a coin. The head may appear on the first toss and the tail on the second *or* the tail may appear on the first toss and the head on the second. **Mutually exclusive events** are those in which the occurrence of any of them excludes the occurrence of the others. The word "or" is usually required or implied in the phrasing of problem solutions involving mutually exclusive events, signaling that an *addition* of probabilities is to be performed. That is, whenever alternative possibilities exist for the satisfaction of the conditions of a problem, the individual probabilities are combined by addition.

Example 2.20 In testcrossing heterozygous black guinea pigs ($Bb \times bb$), there are two ways to obtain one black (Bb) and one white (bb) offspring in a litter of 2 animals. Let p = probability of black = 1/2 and q = probability of white = 1/2.

	First Offspring		Second Offspring		Probability
First alternative:	Black (p)	(and)	White (q)	=	pq
					(or)
Second alternative:	White (q)	(and)	Black (p)	=	qp
			Combined probability	=	$2pq$

$p = q = \frac{1}{2}$; hence the combined probability $= 2(\frac{1}{2})(\frac{1}{2}) = \frac{1}{2}$.

Chapter 3
Two or More Genes

In This Chapter:

- ✔ *Independent Assortment*
- ✔ *Systems for Solving Dihybrid Crosses*
- ✔ *Modified Dihybrid Ratios*
- ✔ *Higher Combinations*

Independent Assortment

In this chapter, we shall consider simultaneously two or more traits, each specified by a different pair of independently assorting autosomal genes, i.e., genes on different chromosomes other than the sex chromosomes.

Example 3.1 In addition to the coat color locus of guinea pigs introduced in Chapter Two (*B-* = black, *bb* = white), another locus on a different chromosome (independently assorting) is known to govern length of hair, such that *L-* = short hair and *ll* = long hair. Any of four different genotypes exist for the black, short-haired phenotype: *BBLL*, *BBLl*, *BbLL*, *BbLl*. Two different genotypes produce a black, long-haired pig: *BBll* or *Bbll*; likewise two genotypes for a white, short-haired pig: *bbLL* or *bbLl*; and only one genotype specifies a white, long-haired pig: *bbll*.

A **dihybrid** genotype is heterozygous at two loci. Dihybrids form four genetically different gametes with approximately equal frequencies because of the random orientation of nonhomologous chromosome pairs on the first meiotic metaphase plate.

Example 3.2 A dihybrid black, short-haired guinea pig (*BbLl*) produces four types of gametes in equal frequencies.

			Gametes	Frequency
B	*L*	=	*BL*	$\frac{1}{4}$
	l	=	*Bl*	$\frac{1}{4}$
b	*L*	=	*bL*	$\frac{1}{4}$
	l	=	*bl*	$\frac{1}{4}$

A summary of the gametic output for all nine genotypes involving two pairs of independently assorting factors is shown below.

Genotypes	Gametes in Relative Frequencies
BBLL	All *BL*
BBLl	½ *BL*: ½ *Bl*
BBll	All *Bl*
BbLL	½ *BL*: ½ *bL*
BbLl	¼ *BL*: ¼ *Bl*: ¼ *bL*: ¼ *bl*
Bbll	½ *Bl*: ½ *bl*
bbLL	All *bL*
bbLl	½ *bL*: ½ *bl*
bbll	All *bl*

A testcross is the mating of an incompletely known genotype to a genotype that is homozygous recessive at all of the loci under consideration. The phenotypes of the offspring produced by a testcross reveal

the number of different gametes formed by the parental genotype under test. When all of the gametes of an individual are known, the genotype of that individual also becomes known. A monohybrid testcross gives a 1:1 phenotypic ratio, indicating that one pair of factors is segregating. A dihybrid testcross gives a 1:1:1:1 ratio, indicating that two pairs of factors are segregating and assorting independently.

Example 3.3 Testcrossing a dihybrid yields a 1:1:1:1 genotypic and phenotypic ratio among the progeny.

Parents:	*BbLl* black, short-haired	×	*bbll* white, long-haired
F_1:	½ *BbLl*		black, short-haired
	½ *Bbll*		black, long-haired
	½ *bbLl*		white, short-haired
	½ *bbll*		white, long-haired

Systems for Solving Dihybrid Crosses

Gametic Checkerboard Method

When two dihybrids are crossed, four kinds of gametes are produced in equal frequencies in both the male and the female. A 4 × 4 gametic checkerboard can be used to show all sixteen possible combinations of these gametes. This method is laborious and time-consuming, and offers more opportunities for error than the other methods that follow.

Example 3.4 P: *BBLL* black, short × *bbll* white, long

F_1: *BbLl* = black, short

F_2:

		Male Gametes			
		BL	*Bl*	*bL*	*bl*
Female Gametes	*BL*	*BBLL* black short	*BBLl* black short	*BbLL* black short	*BbLl* black short
	Bl	*BBLl* black short	*BBll* black long	*BbLl* black short	*Bbll* black long
	bL	*BbLL* black short	*BbLl* black short	*bbLL* white short	*bbLl* white short
	bl	*BbLl* black short	*Bbll* black long	*bbLl* white short	*bbll* white long

F_2 Summary: **Proportions** **Genotypes**

Proportions	Genotypes
$\frac{1}{16}$	*BBLL*
$\frac{1}{8}$	*BBLl*
$\frac{1}{16}$	*BBll*
$\frac{1}{8}$	*BbLL*
$\frac{1}{4}$	*BbLl*
$\frac{1}{8}$	*Bbll*

$\frac{1}{16}$	*bbLL*
$\frac{1}{8}$	*bbLl*
$\frac{1}{16}$	*bbll*

Proportions	**Genotypes**
$\frac{9}{16}$	Black, short
$\frac{3}{16}$	Black, long
$\frac{3}{16}$	White, short
$\frac{1}{16}$	White, long

Genotypic and Phenotypic Checkerboard Methods

A knowledge of the monohybrid probabilities presented in Chapter Two may be applied in a simplified genotypic or phenotypic checkerboard.

Example 3.5 Genotypic checkerboard.

F_1: *BbLl* × *BbLl*
black, short black, short

Considering only the B locus, *Bb* × *Bb* produces ¼ *BB*, ½ *Bb*, and ¼ *bb*. Likewise for the L locus, *Ll* × *Ll* produces ¼ *LL*, ½ *Ll*, and ¼ *ll*. Let us place these genotypic probabilities in a checkerboard and combine independent probabilities by multiplication.

F_2:

	¼*LL*	½*Ll*	¼*ll*
¼*BB*	$\frac{1}{16}$*BBLL*	⅛*BBLL*	$\frac{1}{16}$*BBll*
½*Bb*	⅛*BbLL*	¼*BbLl*	⅛*Bbll*
¼*bb*	$\frac{1}{16}$*bbLL*	⅛*bbLl*	$\frac{1}{16}$*bbll*

Example 3.6 Phenotypic checkerboard.

F_1: *BbLl* × *BbLl*
black, short black, short

Considering the B locus, *Bb* × *Bb* produces ¾ black and ¼ white. Likewise at the L locus, *Ll* × *Ll* produces ¾ short and ¼ long. Let us place these phenotypic probabilities in a checkerboard and combine them by multiplication.

F_2:

	¾ Black	¼ White
¾ Short	$\frac{9}{16}$ Black, short	$\frac{3}{16}$ White, short
¼ Long	$\frac{3}{16}$ Black, long	$\frac{1}{16}$ White, long

Branching Systems

This procedure is used to determine all possible ways in which any number of chromosome pairs could orient themselves on the first meiotic metaphase plate. It can also be used to find all possible genotypic or phenotypic combinations.

Example 3.7 Genotypic trichotomy.

			Ratio	Genotypes
$\frac{1}{4}$ *BB*	$\frac{1}{4}$ *LL*	=	$\frac{1}{16}$	*BB LL*
	$\frac{1}{2}$ *Ll*	=	$\frac{1}{8}$	*BB Ll*
	$\frac{1}{4}$ *ll*	=	$\frac{1}{16}$	*BB ll*
$\frac{1}{2}$ *Bb*	$\frac{1}{4}$ *LL*	=	$\frac{1}{8}$	*Bb LL*
	$\frac{1}{2}$ *Ll*	=	$\frac{1}{4}$	*Bb Ll*
	$\frac{1}{4}$ *ll*	=	$\frac{1}{8}$	*Bb ll*
$\frac{1}{4}$ *bb*	$\frac{1}{4}$ *LL*	=	$\frac{1}{16}$	*bb LL*
	$\frac{1}{4}$ *Ll*	=	$\frac{1}{8}$	*bb Ll*
	$\frac{1}{4}$ *ll*	=	$\frac{1}{16}$	*bb ll*

Example 3.8 Phenotypic dichotomy.

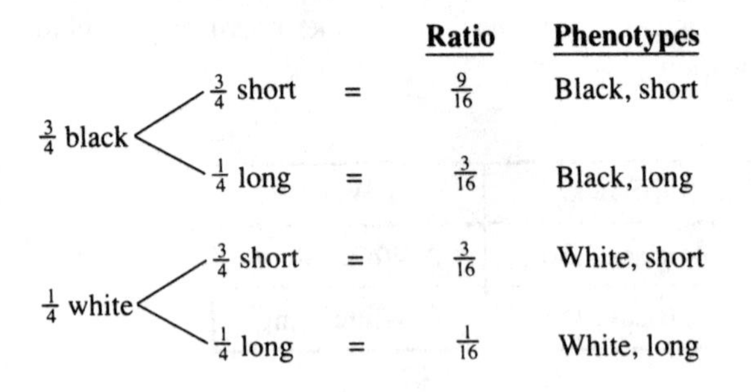

If only one of the genotypic frequencies or phenotypic frequencies is required, there is no need to be concerned with any other genotypes or phenotypes. A mathematical solution can be readily obtained by combining independent probabilities.

Example 3.9 To find the frequency of white, short pigs in the offspring of dihybrid parents, first consider each locus separately:

$Bb \times Bb = 1/4$ white (bb); $Ll \times Ll = 3/4$ short (L-).

Combining these independent probabilities, $1/4 \times 3/4 = \frac{3}{16}$ white, short.

Modified Dihybrid Ratios

The classical phenotypic ratio resulting from the mating of dihybrid genotypes is 9 : 3 : 3 : 1. This ratio appears whenever the alleles at both loci display dominant and recessive relationships. The classical dihybrid ratio may be modified if one or both loci have codominant alleles or lethal alleles. A summary of these modified phenotypic ratios in adult progeny is shown below.

Allelic Relationships in Dihybrid Parents		Expected Adult Phenotypic Ratio
First Locus	Second Locus	
Dominant-recessive	Codominants	3 : 6 : 3 : 1 : 2 : 1
Codominants	Codominants	1 : 2 : 1 : 2 : 4 : 2 : 1 : 2 : 1
Dominant-recessive	Codominant lethal*	3 : 1 : 6 : 2
Codominant	Codominant lethal*	1 : 2 : 1 : 2 : 4 : 2
Lethal*	Codominant lethal*	4 : 2 : 2 : 1

Higher Combinations

The methods for solving two-factor crosses may easily be extended to solve problems involving three or more pairs of independently assorting autosomal factors. Given any number of heterozygous pairs of factors (n) in the F_1, the following general formulas apply:

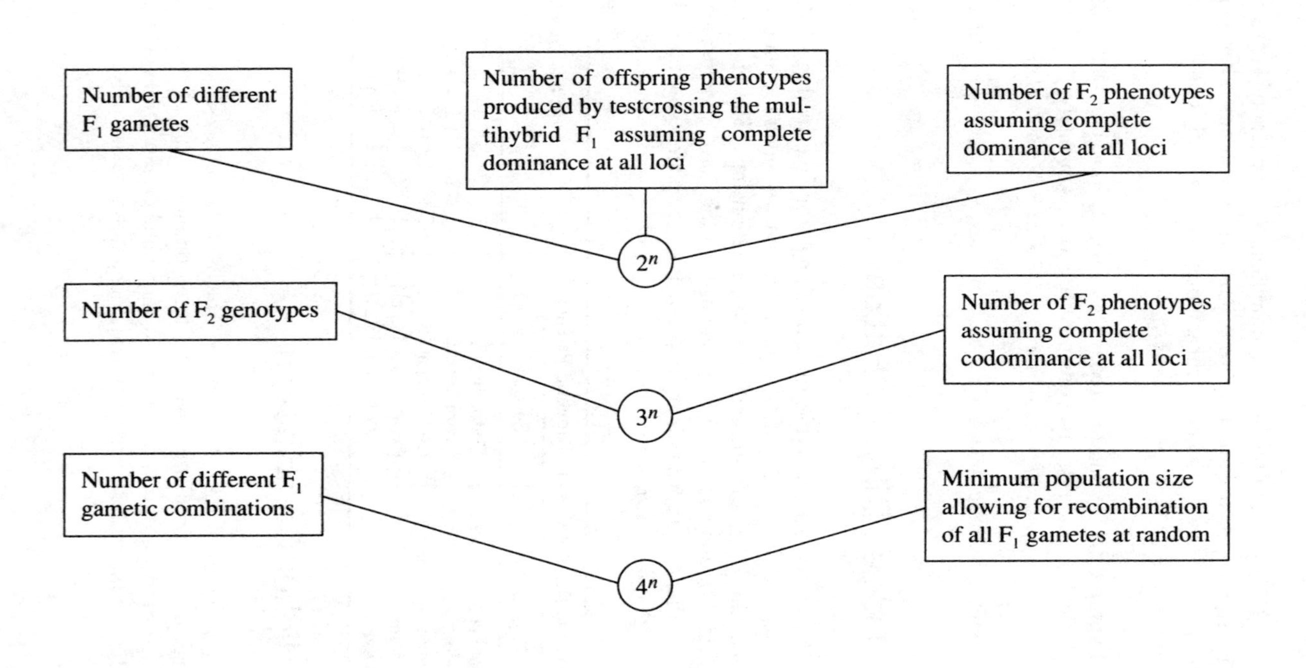
Number of different F_1 gametes
Number of offspring phenotypes produced by testcrossing the multihybrid F_1 assuming complete dominance at all loci
Number of F_2 phenotypes assuming complete dominance at all loci
2^n
Number of F_2 genotypes
Number of F_2 phenotypes assuming complete codominance at all loci
3^n
Number of different F_1 gametic combinations
Minimum population size allowing for recombination of all F_1 gametes at random
4^n

Chapter 4
GENETIC INTERACTION

IN THIS CHAPTER:

- ✔ *Two-Factor Interactions*
- ✔ *Epistatic Interactions*
- ✔ *Nonepistatic Interactions*
- ✔ *Interactions with Three or More Factors*
- ✔ *Pleiotropism*

Two-Factor Interactions

The **phenotype** is a result of gene products brought to expression in a given environment. The environment includes not only external factors such as temperature and the amount or quality of light but also internal factors such as hormones and enzymes. Genes specify the structure of proteins. Most known enzymes are proteins. **Enzymes** perform catalytic functions, causing the splitting or union of various molecules. **Metabolism** is the sum of all the physical and chemical processes by which living protoplasm is produced and maintained and by which energy is made available for the uses of the organism. These biochemical reactions occur as stepwise conversions of one substance into another, each step being mediated by a specific enzyme. All of the steps that transform a precursor substance to its end product constitute a **biosynthetic pathway**.

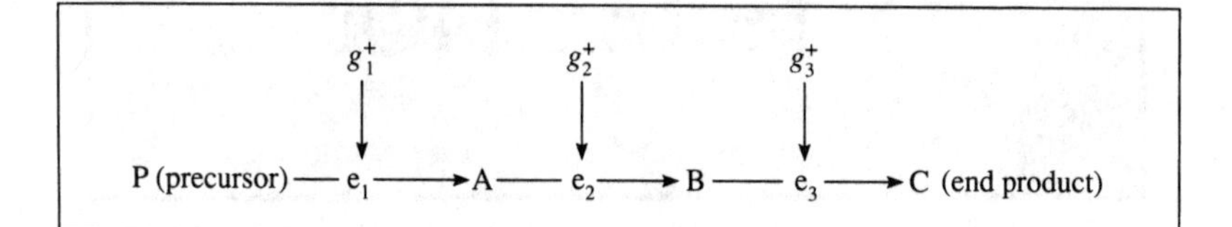

Several genes are usually required to specify the enzymes involved in even the simplest pathways. Each metabolite (A, B, C) is produced by the catalytic action of different enzymes (e_x) specified by different wild-type genes (g_x^+). **Genetic interaction** occurs whenever two or more genes specify enzymes that catalyze steps in a common pathway. If substance C is essential for the production of a normal phenotype, and the recessive mutant alleles g_1, g_2, and g_3 produce defective enzymes, then a mutant (abnormal) phenotype would result from a genotype homozygous recessive at any of three loci. If g_3 is mutant, the conversion of B to C does not occur and substance B tends to accumulate in excessive quantity; if g_2 is mutant, substance A will accumulate. Thus, mutants are said to produce "metabolic blocks."

An organism with a mutation in only g_2 could produce a normal phenotype if it were given either substance B or C, but an organism with a mutation in g_3 has a specific requirement for C. Thus, gene g_3^+ becomes dependent on g_2^+ for its expression as a normal phenotype. If the geneotype is homozygous for the recessive g_2 allele, then the pathway ends with substance A. Neither g_3^+ nor its recessive allele g_3 has any effect on the phenotype. Thus, genotype g_2g_2 can hide or mask the phenotypic expression of alleles at the g_3 locus.

Originally, a gene or locus that suppressed or masked the action of a gene at another locus was termed **epistatic**. The gene or locus suppressed was **hypostatic**. Later, it was found that both loci could be mutually epistatic to one another. Dominance involves *intra*allelic gene suppression, or the masking effect that one allele has upon the expression of another allele at the same locus. Epistasis involves *inter*allelic gene suppression, or the masking effect that one gene locus has upon the expression of another.

You Need to Know ✓

The classical phenotypic ratio of 9 : 3 : 3 : 1 observed in the progeny of dihybrid parents becomes modified by epistasis into ratios that are various combinations of the 9 : 3 : 3 : 1 groupings.

Example 4.1 A particularly illuminating example of gene interaction occurs in white clover. Some strains have a high cyanide content; others have a low cyanide content. Crosses between two strains with low cyanide have produced as F_1 with a high concentration of cyanide in their leaves. The F_2 shows a ratio of 9 high cyanide : 7 low cyanide. Cyanide is known to be produced from the substrate cyanogenic glucoside by enzymatic catalysis. One strain of clover has the enzyme but not the substrate. The other strain makes substrate but is unable to convert it to cyanide. The pathway may be diagrammed as follows where G^x produces an enzyme and g^x results in a metabolic block.

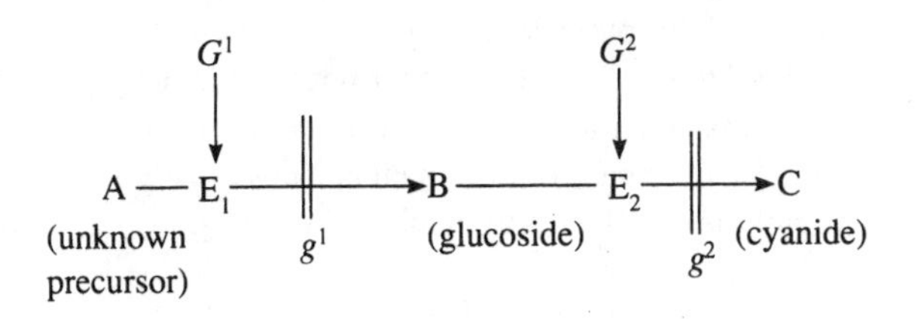

Tests on leaf extracts have been made for cyanide content before and after the addition of either glucoside or the enzyme E_2.

F_2 Ratio	Genotype	Leaf Extract Alone	Leaf Extract Plus Glucoside	Leaf Extract Plus E_2
9	G^1-G^2-	+	+	+
3	G^1-g^2g^2	0	0	+
3	$g^1g^1G^2$-	0	+	0
1	$g^1g^1g^2g^2$	0	0	0

Legend: + = cyanide present, 0 = no cyanide present.

If the leaves are phenotypically classified on the basis of cyanide content of extract alone, a ratio of 9 : 7 results. If the phenotypic classification is based either on extract plus glucoside or on extract plus E_2, a ratio of 12 : 4 is produced. If all of these tests form the basis of phenotypic classification, the classical 9 : 3 : 3 : 1 ratio emerges.

Epistatic Interactions

When epistasis is operative between two gene loci, the number of phenotypes appearing in the offspring from dihybrid parents will be less than four. There are six types of epistatic ratios commonly recognized, three of which have three phenotypes and the other three having only two phenotypes.

Dominant Epistasis (12 : 3 : 1)

When the dominant allele at one locus, for example, the *A* allele, produces a certain phenotype regardless of the allelic condition of the other locus, then the A locus is said to be epistatic to the B locus. Furthermore, since the dominant allele *A* is able to express itself in the presence of either *B* or *b*, this is a case of dominant epistasis. Only when the genotype of the individual is homozygous

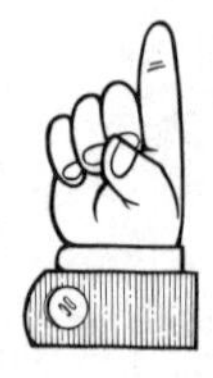

recessive at the epistatic locus (*aa*) can the alleles of the hypostatic locus (*B* or *b*) be expressed. Thus, the genotypes *A-B-* and *A-bb* produce the same phenotype, where as *aaB-* and *aabb* produce two additional phenotypes. The classical 9 : 3 : 3 : 1 ratio becomes modified into a 12 : 3 : 1 ratio.

Recessive Epistasis (9 : 3 : 4)

If the recessive genotype at one locus (e.g., *aa*) suppresses the expression of alleles at the B locus, the A locus is said to exhibit recessive epistasis over the B locus. Only if the dominant allele is present at the A locus can the alleles of the hypostatic B locus be expressed. The genotypes *A-B-* and *A-bb* produce two additional phenotypes. The 9 : 3 : 3 : 1 ratio becomes a 9 : 3 : 4 ratio.

Duplicate Genes with Cumulative Effect (9 : 6 : 1)

If the dominant condition (either homozygous or heterozygous) at either locus (but not both) produces the same phenotype, the F_2 ratio becomes 9 : 6 : 1. For example, where the epistatic genes are involved in producing various amounts of a substance such as pigment, the dominant genotypes of each locus may be considered to produce one unit of pigment independently. Thus, genotypes *A-bb* and *aaB-* produce one unit of pigment each and therefore have the same phenotype. The genotype *aabb* produces no pigment, but in the genotype *A-B-*, the effect is cumulative and two units of pigment are produced.

Duplicate Dominant Genes (15 : 1)

The 9 : 3 : 3 : 1 ratio is modified into a 15 : 1 ratio if the dominant alleles of both loci each produce the same phenotype without cumulative effect.

Duplicate Recessive Genes (9 : 7)

In the case where identical phenotypes are produced by both homozygous recessive genotypes, the F_2 ratio becomes 9 : 7. The genotypes *aaB-*, *A-bb*, and *aabb* produce one phenotype. Both dominant alleles, when present together, complement each other and produce a different phenotype.

Dominant-and-Recessive Interaction (13 : 3)

Only two F_2 phenotypes result when a dominant genotype at one locus (e.g., *A-*) and the recessive genotype at the other (*bb*) produce the same phenotypic effect. Thus, *A-B-*, *A-bb*, and *aabb* produce one phenotype and *aaB-* produces another in the ratio 13 : 3 (see Table 4.1).

Genotypes	*A-B-*	*A-bb*	*aaB-*	*aabb*
Classical ratio	9	3	3	1
Dominant epistasis	12		3	1
Recessive epistasis	9	3	4	
Duplicate genes with cumulative effect	9	6		1
Duplicate dominant genes	15			1
Duplicate recessive genes	9	7		
Dominant and recessive interaction	13		3	

Table 4.1

Nonepistatic Interactions

Genetic interaction may also occur without epistasis if the end products of different pathways each contribute to the same trait.

Example 4.2 The dull-red eye color characteristic of wild-type flies is a mixture of two kinds of pigments (B and D) each produced from non-pigmented compounds (A and C) by the action of different enzymes (e_1 and e_2) specified by different wild-type genes (g_1^+ and g_2^+).

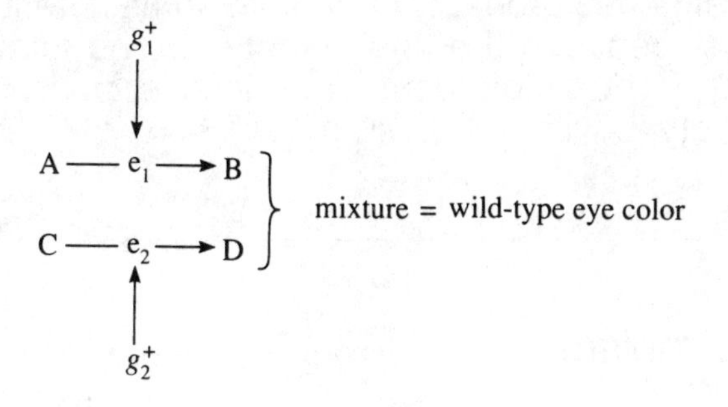

The recessive alleles at these two loci (g_1 and g_2) specify enzymatically inactive proteins. Thus, a genotype without either dominant allele would not produce any pigmented compounds and the eye color would be white.

Phenotypes	Genotypes	End Products
Wild type	$g_1^+/\text{-}$, $g_2^+/\text{-}$	B and D
Color B	$g_1^+/\text{-}$, g_2/g_2	B and C
Color D	g_1/g_1, $g_2^+/\text{-}$	D and A
White	g_1/g_1, g_2/g_2	A and C

Interactions with Three or More Factors

Recall that the progeny from trihybrid parents are expected in the phenotypic ratio 27 : 9 : 9 : 9 : 3 : 3 : 3 : 1. This classical ratio can also be modified whenever two or all three of the loci interact. Interactions involving four or more loci are also possible.

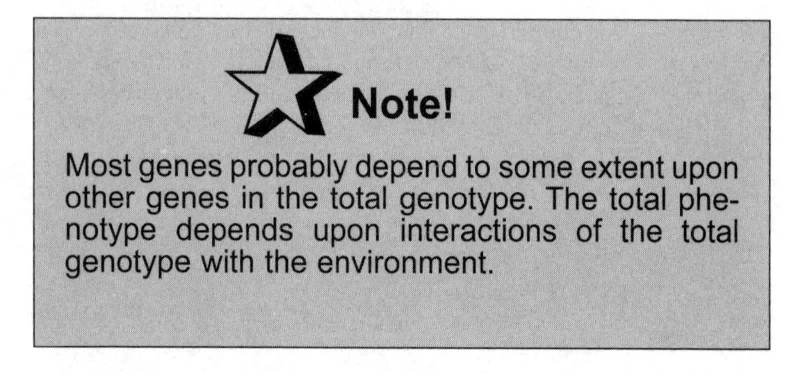

Pleiotropism

Many and perhaps most of the biochemical pathways in the living organism are interconnected and often interdependent. Products of one reaction chain may be used in several other metabolic schemes. It is not surprising, therefore, that the phenotypic expression of a gene usually involves more than one trait. Sometimes one trait will be clearly evident (major effect) and other, perhaps seemingly unrelated ramifications (secondary effects) will be less evident to the casual observer. In other cases, a number of related changes may be considered together as a **syndrome**. All of the manifold phenotypic expressions of a single gene are spoken of as **pleiotropic** gene effects.

Example 4.3 The syndrome called "sickle-cell anemia" in humans is due to an abnormal hemoglobin. This is the primary effect of the mutant gene. Subsidiary effects of the abnormal hemoglobin include the sickle shape of the cells and their tendency to clump together and clog blood vessels in various organs of the body. As a result, damage to the heart, kidney, spleen, and brain are common elements of the syndrome. Defective corpuscles are readily destroyed in the body, causing severe anemia.

Chapter 5
The Genetics of Sex

In This Chapter:

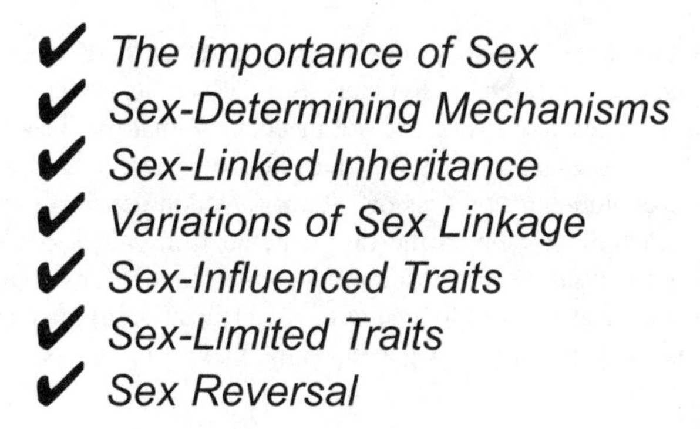

- *The Importance of Sex*
- *Sex-Determining Mechanisms*
- *Sex-Linked Inheritance*
- *Variations of Sex Linkage*
- *Sex-Influenced Traits*
- *Sex-Limited Traits*
- *Sex Reversal*

The Importance of Sex

We are probably too accustomed to thinking of sex in terms of the males and females of our own domestic species. Plants also have sexes; at least we know that there are male and female portions of a flower. Although some of the lowest forms of animal life may have several sexes, the number of sexes is only two in most higher organisms. These sexes may reside in different individuals or within the same individual. An animal possessing both male and female reproductive organs is usually referred to as a **hermaphrodite**. In plants where **staminate** (male) and **pistillate** (female) flowers occur on the same plant, the term of preference is **monoecious**. Moreover, most flowering plants have both male and female parts within the same flower (perfect flower). Relatively few angiosperms are **diecious**, i.e., having the male and female elements in different individuals.

Among the common cultivated crops known to be diecious are asparagus, date palm, hemp, hops, and spinach.

Whether there are two or more sexes, or whether these sexes reside in the same or different individuals is relatively unimportant. The importance of sex itself is that it is a mechanism that provides for the great amount of genetic variability characterizing most natural populations. The evolutionary process of natural selection depends upon this genetic variability to supply the raw material from which the better-adapted types usually survive to reproduce their kind. Many subsidiary mechanisms have evolved to ensure cross-fertilization in most species as a means for generating new genetic combinations in each generation.

Sex-Determining Mechanisms

Most mechanisms for the determination of sex are under genetic control and may be classified into one of the following categories.

Sex Chromosome Mechanisms

Heterogametic Males. In humans, and apparently in all other mammals, the presence of the Y chromosome may determine a tendency to maleness. Normal males are chromosomally XY and females are XX. This produces a 1 : 1 sex ratio in each generation. Since the male produces two kinds of gametes as far as the sex chromosomes are concerned, he is said to be the **heterogametic** sex. The female, producing only one kind of gamete, is the **homogametic** sex. This mode of sex determination is commonly referred to as the XY method.

Example 5.1 XY Method of Sex Determination

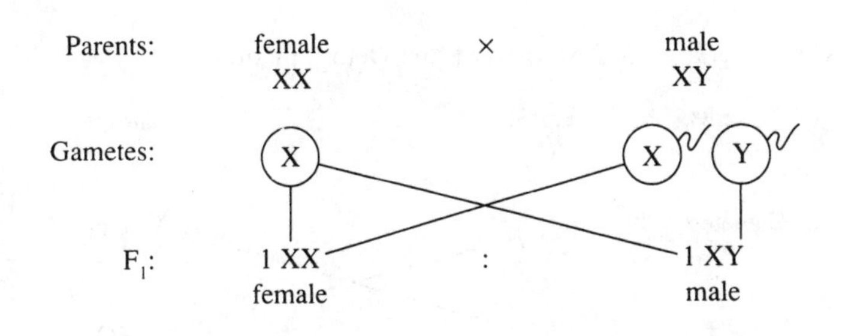

A **testis-determining factor** (TDF) has been delimited to a small segment on the short arm of the human Y chromosome. It is suggested that the gene (or possibly a group of closely linked genes) of the TDF produces a DNA-binding protein that activates one or more other genes (probably on different chromosomes) in a hierarchy or cascade of gene activation that governs the development of the testes. In the absence of TDF, rudimentary gonadal tissue of the embryo would normally develop into an ovary. TDF seems to be highly conserved in mammals. The location of TDF was aided by the discovery of rare exceptions to the rule that XX programs for femaleness and XY programs for maleness. It was found that normal-appearing but sterile XX human males have at least some of the TDF attached to one of their X chromosomes and human XY females have lost a crucial part of the TDF from their Y chromosome.

In some insects, especially those of the orders Hemiptera (true bugs) and Orthoptera (grasshoppers and roaches), males are also heterogametic, but produce either X-bearing sperm or gametes without a sex chromosome. In males of these species, the X chromosome has no homologous pairing partner because there is no Y chromosome present. Thus, males exhibit an odd number in their chromosome complement. The one-X and two-X condition determines maleness and femaleness, respectively. If the single X chromosome of the male is always included in one of the two types of gametes formed, then a 1 : 1 sex ratio will be produced in the progeny. This mode of sex determination is commonly referred to as the XO method where the O symbolizes the lack of a chromosome analogous to the Y of the XY system.

Example 5.2 XO Method of Sex Determination

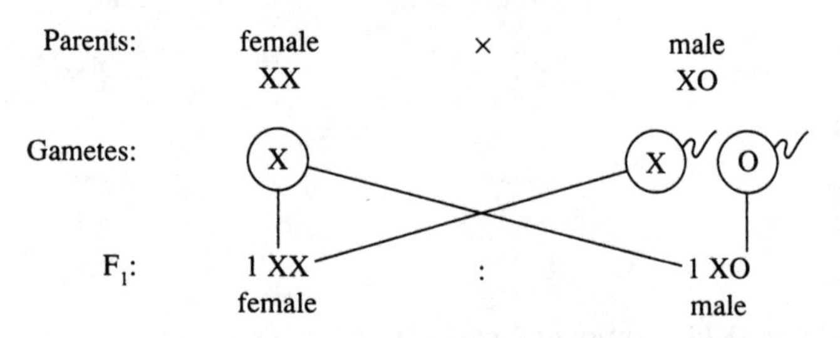

Heterogametic Females. This method of sex determination is found in a comparatively large group of insects including the butterflies, moths, caddis flies, and silkworms, and in some birds and fish. The 1-X and 2-X condition in these species (e.g., domestic chickens) have a chromosome similar to that of the Y in humans. In these cases, the chromosomes are sometimes labeled Z and W instead of X and Y, respectively, in order to call attention to the fact that the female (ZW) is the heterogametic sex and the male (ZZ) is the homogametic sex. The females of other species have no homologue to the single sex chromosome as in the case of the XO mechanism discussed previously. To point out this difference, the symbols ZZ and ZO may be used to designate males and females, respectively. A 1 : 1 sex ratio is expected in either case.

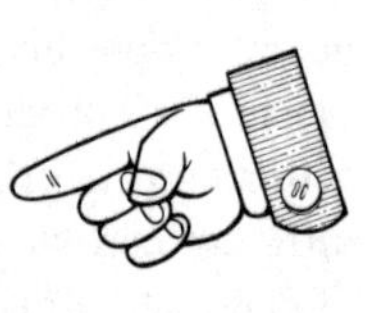

Example 5.3 ZO Method of Sex Determination

Example 5.4 ZW Method of Sex Determination

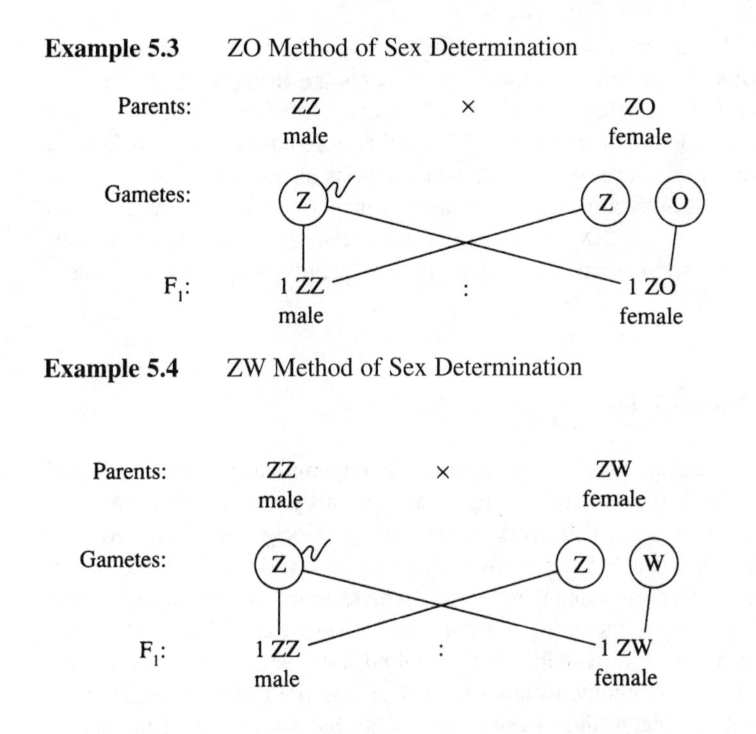

The W chromosome of the chicken is not a strong female-determining element. Recent studies indicate that sex determination in chickens, and probably birds in general, is similar to that of *Drosophila*, i.e., dependent upon the ratio between the Z chromosomes and the number of autosomal sets of chromosomes.

Genic Balance

The presence of the Y chromosome in *Drosophila*, although it is essential for male fertility, apparently has nothing to do with the determination of sex. Instead, the factors for maleness residing in all of the autosomes are "weighed" against the factors for femaleness residing on the X chromosome(s). If each haploid set of autosomes carries factors with a male-determining value equal to 1, then each X chromosome carries factors with a female-determining value of 1½.

Let A represent a haploid set of autosomes. In a normal male (AAXY), the male : female determinants are in the ratio 2 : 1½ and therefore the balance is in favor of maleness. A normal female (AAXX) has a male : female ratio of 2 : 3 and therefore the balance is in favor of femaleness. Several abnormal combinations of chromosomes have confirmed this hypothesis. For example, an individual with three sets of autosomes and 2 X chromosomes has a ratio of 3 : 3, which makes its genetic sex neutral, and indeed phenotypically it appears as a sterile intersex.

Haplodiploidy

Male bees are known to develop **parthenogenetically** (without union of gametes) from unfertilized eggs (**arrhenotoky**) and are therefore haploid. Females (both workers and queens) originate from fertilized (diploid) eggs. Sex chromosomes are not involved in this mechanism of sex determination, which is characteristic of the insect order Hymenoptera including the ants, bees, wasps, etc. The quantity and quality of food available to the diploid larva determines whether that female will become a sterile worker or a fertile queen. Thus, environment here determines sterility or fertility but does not alter the genetically determined sex. The sex ratio of the offspring is under the control of the queen. Most of the eggs laid in the hive will be fertilized and develop into worker females. Those eggs which the queen chooses not to fertilize (from her store of sperm in the seminal receptacle) will develop into fertile haploid males. Queen bees usually mate only once during their lifetime.

Single-Gene Effects

Complementary Sex Factors. At least two members of the insect order Hymenoptera are known to produce males by homozygosity at a single-gene locus as well as by haploidy. This has been confirmed in the tiny parasitic wasp *Bracon hebetor* (often called *Habrobracon juglandis*), and more recently, in bees also. At least nine sex alleles are known at this locus in *Bracon* and may be represented by s^a, s^b, s^c, …, s^i. All females must be heterozygotes such as $s^a s^b$, $s^a s^c$, $s^d s^f$, etc. If an individual is homozygous for any of these alleles such as $s^a s^a$, $s^c s^c$, etc, it develops into a diploid male (usually sterile). Haploid males, of course, would carry only one of the alleles at this locus, e.g., s^a, s^c, s^g, etc.

Example 5.5

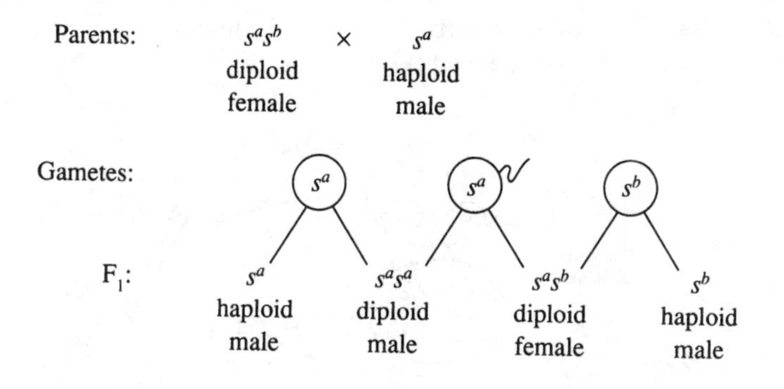

Among the diploid progeny, we expect 1 $s^a s^a$ male : 1 $s^a s^b$ female. Among the haploid progeny, we expect 1 s^a male : 1 s^b male.

The "Transformer" Gene of *Drosophila*. A recessive gene (*tra*) on chromosome 3 of *Drosophila*, when homozygous, transforms a diploid female into a sterile male. The X/X, *tra/tra* individuals resemble normal males in external and internal morphology with the exception that the testes are much reduced in size. The gene is without effect in normal males. The presence of this gene can considerably alter the sex ratio.

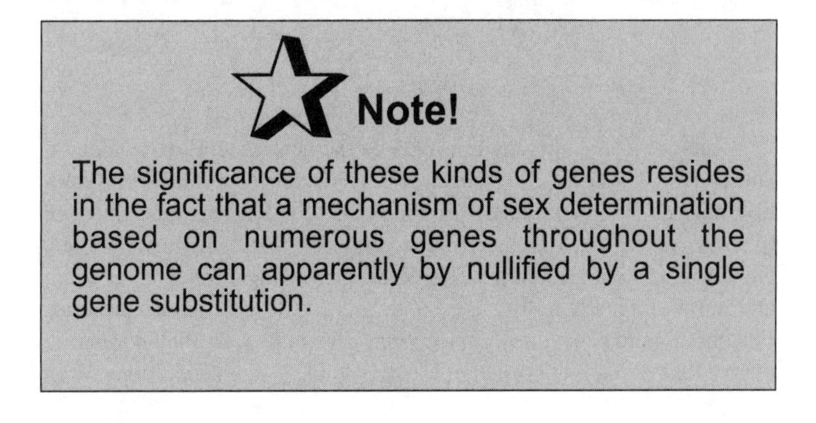

The significance of these kinds of genes resides in the fact that a mechanism of sex determination based on numerous genes throughout the genome can apparently by nullified by a single gene substitution.

"Mating Type" in Microorganisms. In micororganisms such as the alga *Chlamydomonas* and the fungi *Neurospora* and yeast, sex is under control of a single gene. Haploid individuals possessing the same allele of this "mating-type" locus usually cannot fuse with each other to form a zygote, but haploid cells of opposite (complementary) allelic constitution at this locus may fuse. Asexual reproduction in the single-celled motile alga *Chlamydomonas reinhardi* usually involves two mitotic divisions within the old cell wall (Figure 5-1).

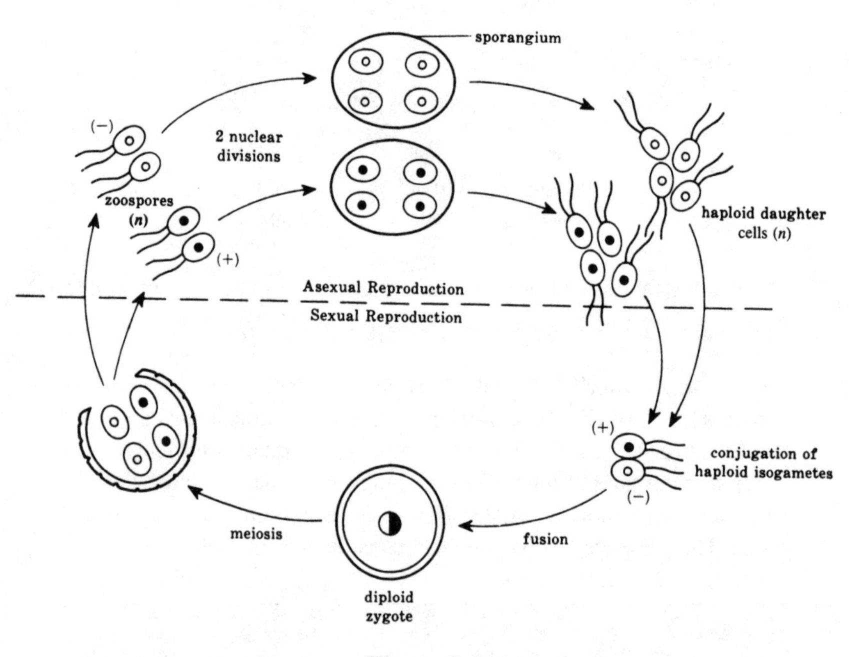

Figure 5-1

Rupture of the sporangium releases the new generation of haploid **zoospores**. If nutritional requirements are satisfied, asexual reproduction may go on indefinitely. In unfavorable conditions where nitrogen balance is upset, daughter cells may be changed to gametes. Genetically, there are two mating types, plus (+) and minus (−), which are morphologically indistinguishable and therefore called **isogametes**. Fusion of gametes unites two entire cells into a diploid nonmotile zygote that is relatively resistant to unfavorable growth conditions. With the return of conditions favoring growth, the zygote experiences meiosis and forms four motile haploid daughter cells (zoospores), two of plus and two of minus mating type.

Sex-Linked Inheritance

Any gene located on the X chromosome (mammals, *Drosophila*, and others) or on the analogous Z chromosome (in birds and other species with the ZO or ZW mechanism of sex determination) is said to be **sex-linked**. The first sex-linked gene found in *Drosophila* was the recessive white-eye mutation. Reciprocal crosses involving autosomal traits yield comparable results. This is not true with sex-linked traits, as shown below. When white-eyed females are crossed with wild-type (red-eyed) males, all the male offspring have white eyes like their mother and all the female offspring have red eyes like their father.

Example 5.6

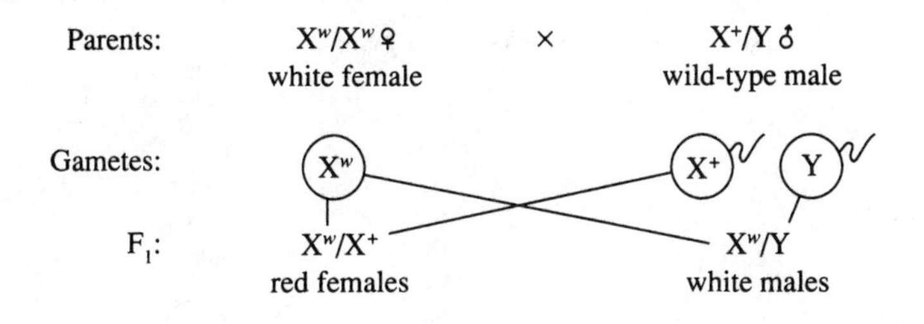

This crisscross method of inheritance is characteristic of sex-linked genes. This peculiar type of inheritance is due to the fact that the Y chromosome carries no alleles homologous to those at the white locus on the X chromosome. In fact, in most organisms with the Y-type chromosome, the Y is virtually devoid of known genes. Thus, males carry only one allele for sex-linked traits. This one-allelic condition is termed **hemizygous** in contrast to the homozygous or heterozygous possibilities in the female. If the F_1 of Example 5.6 mate among themselves to produce an F_2, a 1 red : 1 white phenotypic ratio is expected in both the males and females.

Example 5.7

F_1: X^+/X^w red female × X^w/Y white male

F_2:

	X^+	X^w
X^w	X^+/X^w red female	X^w/X^w white female
Y	X^+/Y red male	X^w/Y white male

The reciprocal cross, where the sex-linked mutation appears in the male parent, results in the disappearance of the trait in the F_1 and its reappearance only in the males of the F_2. This type of skip-generation inheritance also characterizes sex-linked genes.

Example 5.8

Parents: X^+/X^+ red female × X^w/Y white male

Gametes: X^+ ; X^w, Y

F_1: X^+/X^w red female ; X^+/Y red male

F_2:

	X^+	X^w
X^+	X^+/X^+ red female	X^+/X^w red (carrier) female
Y	X^+/Y red male	X^w/Y white male

Thus, a 3 red : 1 white phenotypic ratio is expected in the total F_2 disregarding sex, but only the males show the mutant trait. The phenotypic ratio among the F_2 males is 1 red : 1 white. All F_2 females are phenotypically wild-type.

In normal diploid organisms with sex-determining mechanisms like that of humans or *Drosophila*, a trait governed by a sex-linked recessive gene usually manifests itself in the following manner: (1) it is usually found more frequently in the male than in the female of the species, (2) it fails to appear in females unless it also appeared in the paternal parent, and (3) it seldom appears in both father and son, then only if the maternal parent is heterozygous. On the other hand, a trait governed by a sex-linked dominant gene usually manifests itself by (1) being found more frequently in the female than in the male of the species, (2) being found in all female offspring of a male that shows the trait, and (3) failing to be transmitted to any son from a mother that did not exhibit the trait herself.

Variations of Sex Linkage

The sex chromosomes (X and Y) often are of unequal size, shape, and/or staining qualities. The fact that they pair during meiosis is indication that they contain at least some homologous segments. Genes on the homologous segments are said to be **incompletely sex-linked** or **partially sex-linked** and may recombine by crossing over in both sexes just as do the gene loci on homologous autosomes. Special crosses are required to demonstrate the presence of such genes on the X chromosome, and few examples are known.

Genes on the nonhomologous segment of the X chromosome are said to be **completely sex-linked** and exhibit the peculiar mode of inheritance described in the preceding sections. In humans, a few genes are known to reside in the nonhomologous portion of the Y chromosome. In such cases, the trait would be expressed only in males and would always be transmitted from father to son. Such completely Y-linked genes are called **holandric genes** (Figure 5-2).

Figure 5-2 Generalized diagram of X and Y chromosomes. The relative size of these chromosomes and the size of homologous and nonhomologous regions, as well as location of the centromeres (not shown) vary according to the species.

X

Nonhomologous portion or differential segment contains completely sex-linked genes

Homologous portions of the X and Y contain incompletely sex-linked genes

Y

Differential segment of the Y contains holandric genes

Sex-Influenced Traits

The genes governing sex-influenced traits may reside on any of the autosomes or on the homologous portions of the sex chromosomes. The expression of dominance or recessiveness by the alleles of sex-influenced loci is reversed in males and females due, in large part, to the difference in the internal environment provided by the sex hormones.

Remember

Examples of sex-influenced traits are most readily found in the higher animals with well-developed endocrine systems.

Example 5.9 The gene for pattern baldness in humans exhibits dominance in men, but acts recessively in women.

Genotypes	Phenotypes	
	Men	Women
$b'b'$	Bald	Bald
$b'b$	Bald	Nonbald
bb	Nonbald	Nonbald

Sex-Limited Traits

Some autosomal genes may only come to expression in one of the sexes either because of differences in the internal hormonal environment or because of anatomical dissimilarities. For example, we know that bulls have many genes for milk production that they may transmit to their daughters, but they or their sons are unable to express this trait. The production of milk is therefore limited to variable expression in only the female sex. When the penetrance of a gene in one sex is zero, the trait will be **sex-limited**.

Example 5.10 Chickens have a recessive gene for cock-feathering that is penetrant only in the male environment.

Genotype	Phenotypes	
	Males	Females
HH	Hen-feathering	Hen-feathering
Hh	Hen-feathering	Hen-feathering
hh	Cock-feathering	Hen-feathering

Sex Reversal

Female chickens (ZW) that have laid eggs have been known to undergo not only a reversal of the secondary sexual characteristics such as development of cock-feathering, spurs, and crowing, but also the development of testes and even the production of sperm cells (primary sexual characteristics). This may occur when, for example, disease destroys the ovarian tissue, and in the absence of the female sex hormones, the rudimentary testicular tissue present in the center of the ovary is allowed to proliferate. In solving problems involving sex reversals, it must be remembered that the functional male derived through sex reversal will still remain genetically female (ZW).

Solved Problem 5.1. There is a dominant sex-linked gene *B* that places white bars on an adult black chicken as in the Barred Plymouth Rock breed. Newly hatched chicks, which will become barred later in life, exhibit a white spot on the top of the head. (a) Diagram the cross through the F_2 between a homozygous barred male and a nonbarred female. (b) Diagram the reciprocal cross through the F_2 between a homozygous nonbarred male and a barred female. (c) Will both of the above crosses be useful in sexing F_1 chicks at hatching?

Solution.

(a)

Parents: Z^B/Z^B barred male × Z^b/W nonbarred female

Gametes: (*B*) ♂ ; (*b*) (W)

F_1: *B*/*b* barred male ; *B*/W barred female

F_2:

	(*B*)	(W)
(*B*) ♂	*B*/*B* barred male	*B*/W barred female
(*b*) ♂	*B*/*b* barred male	*b*/W nonbarred female

(b)

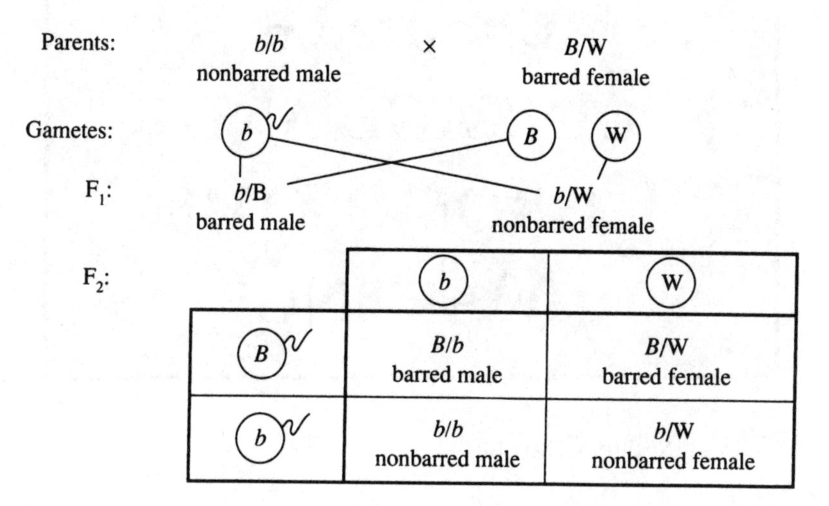

(c) No. Only the cross shown in (b) would be diagnostic in sexing F_1 chicks at birth through the use of this genetic marker. Only male chicks will have a light spot on their heads.

Chapter 6
LINKAGE AND CHROMOSOME MAPPING

IN THIS CHAPTER:

- ✔ *Recombination Among Linked Genes*
- ✔ *Genetic Mapping*
- ✔ *Linkage Estimates from F_2 Data*
- ✔ *Use of Genetic Maps*
- ✔ *Crossover Suppression*

Recombination Among Linked Genes

Linkage

When two or more genes reside in the same chromosome, they are said to be linked. They may be linked together on one of the autosomes or connected together on the sex chromosome. Genes on different chromosomes are distributed into gametes independently of one another (Mendel's Law of Independent Assortment). Genes on the same chromosome, however, tend to stay together during the formation of gametes. Thus, the results of testcrossing dihybrid individuals will yield different results, depending upon whether the genes are linked or on different chromosomes.

Example 6.1 Genes on different chromosomes assort independently, giving a 1: 1: 1: 1 testcross ratio.

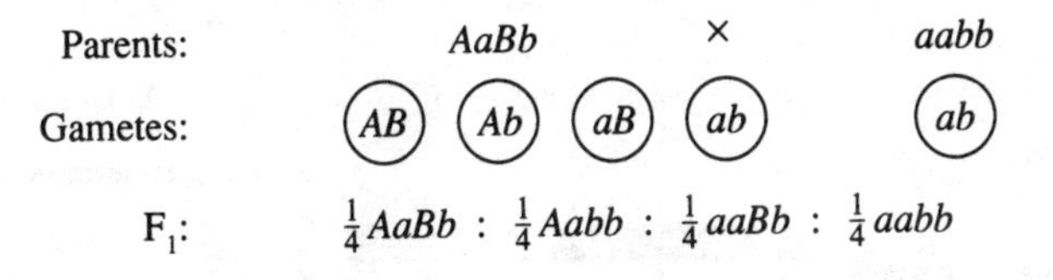

Large deviations from a 1: 1: 1: 1 ratio on the testcross progeny of a dihybrid could be used as evidence for linkage. Linked genes do not always stay together, however, because homologous nonsister chromatids may exchange segments of varying length with one another during meiotic prophase.

> Recall that homologous chromosomes pair with one another in a process called "synapsis" and that the points of genetic exchange, called "chiasmata," produce recombinant gametes through crossing over.

Crossing Over

During meiosis, each chromosome replicates, forming two identical sister chromatids. Homologous chromosomes pair (synapse) and crossing over occurs between nonsister chromatids. This latter process involves the breakage and reunion of only two of the four strands at any given point on the chromosomes. In the diagram below, a crossover occurs in the region between the *A* and *B* loci.

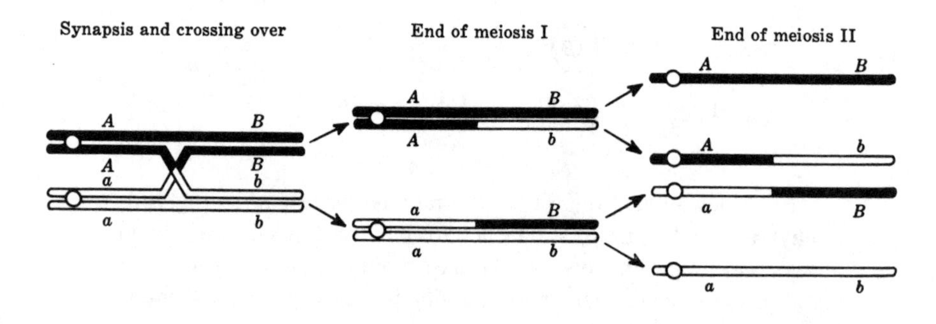

Notice that two of the meiotic products (*AB* and *ab*) have the genes linked in the same way as they were in the parental chromosomes. These products are produced from chromatids that were not involved in crossing over and are referred to as **noncrossover** or **parental** types. The other two meiotic products (*Ab* and *aB*) produced by crossing over have recombined the original linkage relationships of the parent into two new forms called **recombinant** or **crossover** types.

The alleles of double heterozygotes (dihybrids) at two-linked loci may appear in either of two positions relative to one another. If the two dominant (or wild-type) alleles are on one chromsome and the two recessives (or mutants) on the other (AB/ab), the linkage relationship is called **coupling phase**. When the dominant allele of one locus and the recessive allele of the other occupy the same chromosome (*Ab*/*aB*), the relationship is termed **repulsion phase**. Parental and recombinant gametes will be of different types, depending upon how these genes are linked in the parent.

Chiasma Frequency

A pair of synapsed chromosome (bivalent) consists of four chromatids called a *tetrad*. Every tetrad usually experiences at least one chiasma somewhere along its length.

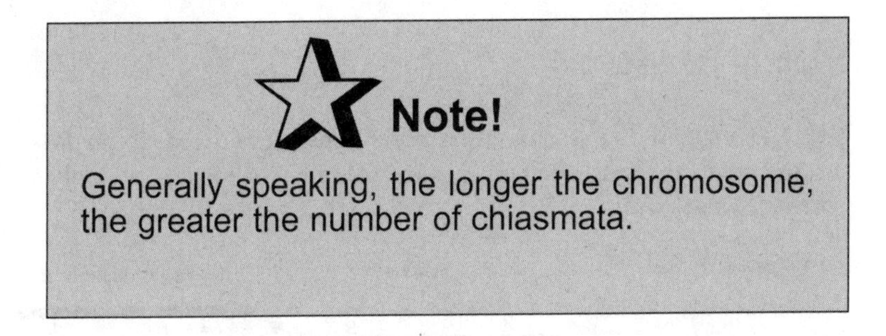

The frequency with which a chiasma occurs between any two genetic loci also has a characteristic or average probability. The further apart two genes are located on a chromosome, the greater the opportunity for a chiasma to occur between them. The closer two genes are linked, the smaller the chance for a chiasma occurring between them. These chiasmata probabilities are useful in predicting the proportions of parental and recombinant gametes expected to be formed from a given genotype. The percentage of crossover (recombinant) gametes formed by a given genotype is a direct reflection of the frequency with which a chiasma forms between the genes in question. Only when a crossover forms *between* the gene loci under consideration will recombination be detected.

Example 6.2 Crossing over outside the A-B region fails to recombine these markers.

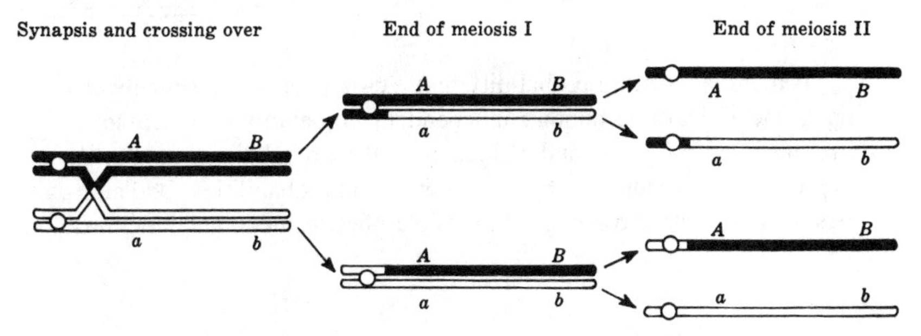

When a chiasma forms between two gene loci, only half of the meiotic products will be of crossover type. Therefore, chiasma frequency is twice the frequency of crossover products.

$$\text{Chiasma } \% = 2(\text{crossover } \%) \text{ or Crossover } \% = \tfrac{1}{2} (\text{chiasma } \%)$$

Multiple Crossovers

When two-strand double crossovers occur between two genetic markers, the products, as detected through the progeny phenotypes, are only parental types.

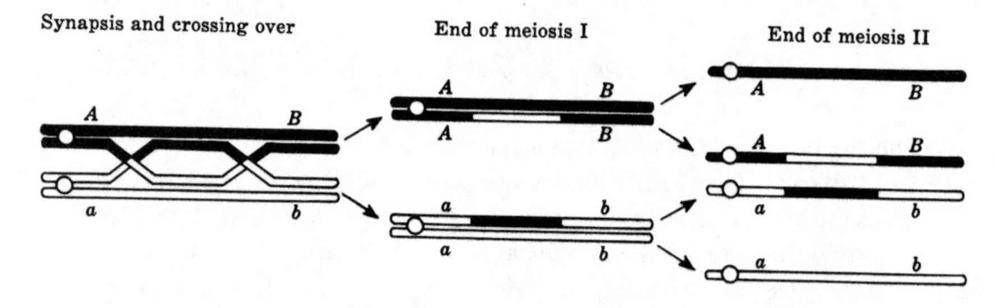

In order to detect these double crossovers, a third gene locus (C) between the outside markers must be used.

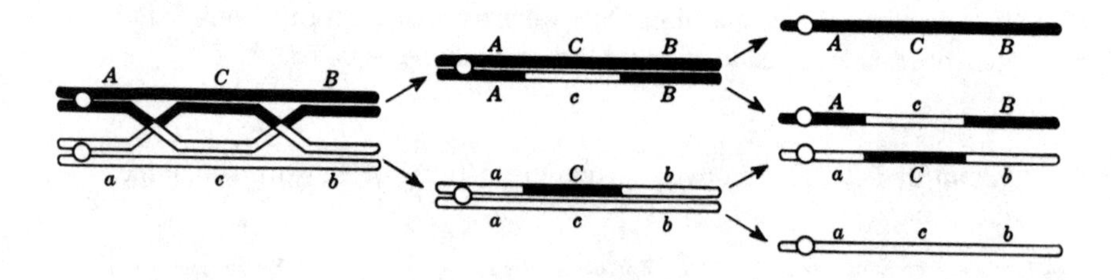

If there is a certain probability that a crossover will form between the A and C loci and another independent probability of a crossover forming between the C and B loci, then the probability of a double crossover is the product of the two independent probabilities. Odd numbers of two-strand crossovers (1, 3, 5, etc.) between two gene loci produce detectable recombinations between the outer markers, but even numbers of two-strand crossovers (2, 4, 6, etc.) do not.

Limits of Recombination

If two gene loci are so far apart in the chromosome that the probability of a chiasma forming between them is 100 percent, then 50 percent of the gametes will be parental type (noncrossover) and 50 percent recombinant (crossover) type. When such dihybrid individuals are testcrossed, they are expected to produce progeny in a 1: 1: 1: 1 ratio as would be expected for genes on different chromosomes. Recombination between two linked genes cannot exceed 50 percent even when multiple crossovers occur between them.

Genetic Mapping

Map Distance

The places where genes reside in the chromosome (loci) are positioned in linear order analogous to beads on a string. There are two major aspects to genetic mapping: (1) the determination of the linear order with which the genetic units are arranged with respect to one another (gene order) and (2) the determination of the relative distances between the genetic units (gene distance). The unit of distance that has the greatest utility in predicting the outcome of certain types of matings is an expression of the probability that crossing over will occur between the two genes under consideration. One unit of map distance (**centimorgan**) is therefore equivalent to 1 percent crossing over.

Each chiasma produces 50 percent crossover products. Fifty percent crossing over is equivalent to 50 map units. If the average (mean) number of chiasmata is known for a chromosome pair, the total length of the map for that linkage may be predicted:

$$\text{Total length} = \text{mean number of chiasmata} \times 50$$

Two-Point Testcross

The easiest way to detect crossover gametes in a dihybrid is through the testcross progeny. Suppose we testcross dihybrid individuals in coupling phase (*AC*/*ac*) and find in the progeny phenotypes 37 percent dominant at both loci, 37 percent recessive at both loci, 13 percent dominant at the first locus and recessive at the second, and 13 percent dominant at the second locus and recessive at the first. Obviously, the last two groups (genotypically *Ac*/*ac* and *aC*/*ac*) were produced by crossover gametes from the dihybrid parent. Thus, 26 percent of all gametes (13 + 13) were of crossover types and the distance between the loci *A* and *C* is estimated to be 26 map units.

Three-Point Testcross

Double crossovers usually do not occur between genes less than five map units apart. For genes further apart, it is advisable to use a third marker between the other two in order to detect any double crossovers. Suppose that we testcross trihybrid individuals of genotype *ABC*/*abc* and find in the progeny the following:

36% *ABC*/*abc*	9% *Abc*/*abc*	4% *ABc*/*abc*	1% *AbC*/*abc*
36% *abc*/*abc*	9% *aBC*/*abc*	4% *abC*/*abc*	1% *aBc*/*abc*
72% Parental type :	18% Single crossovers : between A and B (region I)	8% Single crossovers : between B and C (region II)	2% Double crossovers

To find the distance *A*–*B*, we must count all crossovers (both singles and doubles) that occurred in region I = 18 percent + 2 percent = 20 percent or 20 map units between the loci *A* and *B*. To find the distance *B*–*C*, we must again count all crossovers (both singles and doubles) that occurred in region II = 8 percent + 2 percent = 10 percent or 10 map units between the loci *B* and *C*. The *A*–*C* distance is therefore 30 map units when double crossovers are detected in the two-point linkage experiment above.

Without the middle marker (*B*), double crossovers would appear as parental types and hence we *underestimate* the true map distance (crossover percentage). In this case, the 2 percent double crossovers would appear with the 72 percent parental types, making a total of 74 percent parental types and 26 percent recombinant types. Therefore, for any three linked genes, whose distances are known, the amount of *detectable* crossovers (recombinants) between the two outer markers *A* and *C* when the middle marker *B* is missing is (*A–B* crossover percentage) plus (*B–C* crossover percentage) minus (2 × double-crossover percentage). This procedure is appropriate only if a crossover in the *A–B* region occurs independently of that in the *A–B* region.

Gene Order

The additivity of map distances allows us to place genes in their proper linear order. Three linked genes may be in any one of three different orders, depending upon which gene is in the middle. We will ignore left and right end alternatives for the present. If double crossovers do not occur, map distances may be treated as completely additive units. When we are given the distances *A–B* = 12, *B–C* = 7, and *A–C* = 5, we should be able to determine the correct order.

Case 1. Let us assume that *A* is in the middle.

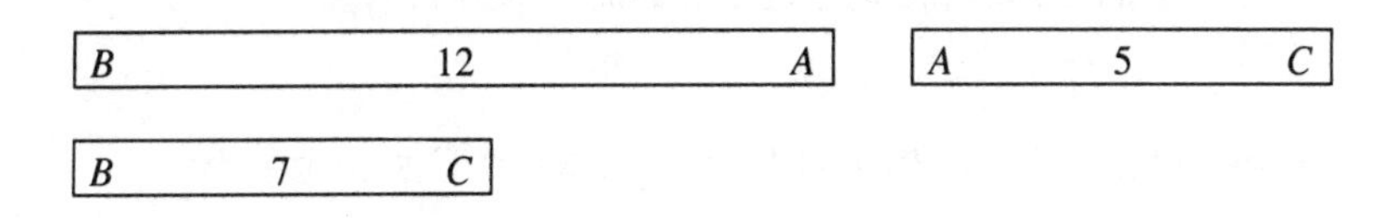

The distance *B–C* is not equitable. Therefore, *A* cannot be in the middle.

Case 2. Let us assume that *B* is in the middle.

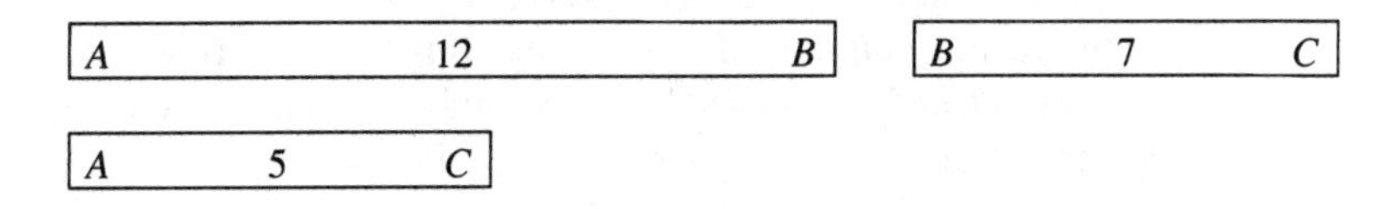

The distance *A–C* is not equitable. Therefore, *B* cannot be in the middle.

Case 3. Let us assume that *C* is in the middle.

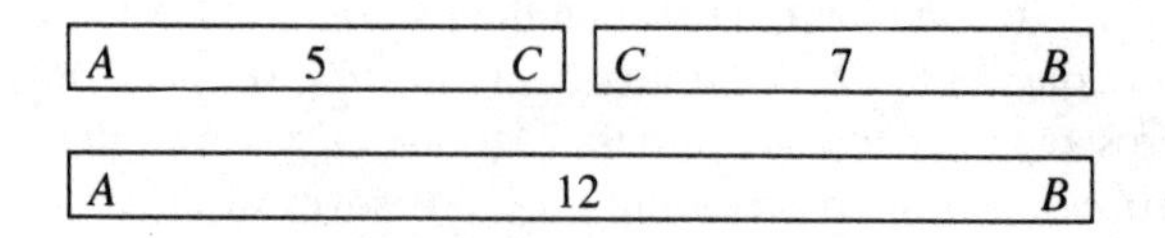

The distance *A–B* is equitable. Therefore, *C* must be in the middle. Most students should be able to perceive the proper relationships intuitively.

Linkage Relationships from a Two-Point Testcross. Parental combinations will tend to stay together in the majority of the progeny and the crossover types will always be the least frequent class. From this information, the mode of linkage (coupling or repulsion) may be determined for the dihybrid parent.

Linkage Relationships from a Three-Point Testcross. In a testcross involving three linked genes, the parental types are expected to be most frequent and the double crossovers to be the least frequent. The gene order is determined by manipulating the parental combinations into the proper order for the production of double-crossover types.

Recombination Percentage vs. Map Distance

In two-point linkage experiments, the chance of double (and other even-numbered) crossovers occurring undetected increases with the unmarked distance (i.e., without segregating loci) between genes. Hence, closely linked genes give the best estimate of crossing over. Double crossovers do not occur within 10-12 map units in *Drosophila.* Minimum double-crossover distance varies by species. Within this minimum distance, recombination percentage is equivalent to map distance. Outside it, the relationship becomes nonlinear (Figure 6-1). True map distance will thus be underestimated by the recombination fraction, with the two becoming virtually independent at large distances.

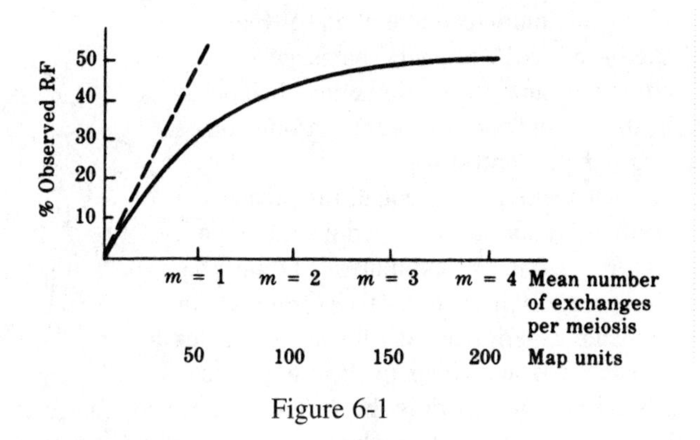

Figure 6-1

Figure 6-1. Relationship between observed recombination frequency (RF) and real map units (solid line). Dashed line represents the relationship for very small mean numbers of exchanges per meiosis (m). From *An Introduction to Genetic Analysis*, 2d ed., by D. T. Suzuki, A. J. F. Griffiths, and R. C. Lewontin, W. H. Freeman and Co., San Francisco, 1976.

Genetic vs. Physical Maps

The frequency of crossing over usually varies in different segments of the chromosome, but is a highly predictable event between any two gene loci. Therefore, the actual physical distances between linked genes bears no direct relationship to the map distances calculated on the basis of crossover percentages. The linear order, however, is identical in both cases.

Combining Map Segments

Segments of map determined from three-point linkage experiments may be combined whenever two of the three genes are held in common.

Additional segments of map added in this manner can produce a total linkage map over 100 map units long. However, as explained previously, the maximum recombination between any two linked genes is 50 percent. That is, genes very far apart on the same chromosome may behave as though they were on different chromosomes (assorting independently).

All other factors being equal, the greater the number of individuals in an experiment, the more accurate the linkage estimates should be. Therefore, in averaging the distances from two or more replicate experiments, the linkage estimates may be **weighted** according to the sample size. For each experiment, multiply the sample size by the linkage estimate. Add the products and divide by the total number of individuals from all experiments.

Interference and Coincidence

In most of the higher organisms, the formation of one chiasma actually reduces the probability of another chiasma forming in an immediately adjacent region of the chromosome. This reduction in chiasma formation may be thought of as being due to a physical inability of the chromatids to bend back upon themselves within certain minimum distances. The net result of this **interference** is the observation of fewer double-crossover types than would be expected according to map distances. The strength of interference varies in different segments of the chromosome and is usually expressed in terms of a **coefficient of coincidence**, or the ratio between the observed and expected double crossovers.

$$\text{Coefficient of coincidence} = \frac{\%\text{ observed double crossovers}}{\%\text{ expected double crossovers}}$$

Coincidence is the complement of interference.

Coincidence + interference = 1.0

When interference is complete (1.0), no double crossovers will be observed and coincidence becomes zero. When we observe all the double crossovers expected, coincidence is unity and interference becomes zero. When interference is 30 percent operative, coincidence becomes 70 percent, etc.

The percentage of double crossovers that will probably be observed can be predicted by multiplying the expected double crossovers by the coefficient of coincidence.

Linkage Estimates from F_2 Data

Sex-Linked Traits

In organisms where the male is XY or XO, the male receives only the Y chromosome from the paternal parent (or no chromosome homologous with the X in the case of XO sex determination). The Y contains, on its differential segment, no alleles homologous to those on the X chromosome received from the maternal parent. Thus, for completely sex-linked traits, the parental and recombinant gametes formed by the female can be observed directly in the F_2 males, regardless of the genotype of the F_1 males.

If the original parental females are double recessive (testcross parent), then both male and female progeny of the F_2 can be used to estimate the percentage of crossing over.

In organisms where the female is the heterogametic sex (ZW or ZO methods of sex determination), the F_2 females can be used for detection of crossing over between sex-linked genes. If the male is used as a testcross parent, both males and females of the F_2 can be used to estimate the strength of the linkage.

Autosomal Traits

A poor alternative to the testcross method for determining linkage and estimating distances is by allowing dihybrid F_1 progeny to produce an

F_2 either by random mating among the F_1 or, in the case of plants, by selfing the F_1. Such an F_2 that obviously does not conform to the 9 : 3 : 3 : 1 ratio expected for genes assorting independently may be considered evidence for linkage. Two methods for estimating the degree of linkage from F_2 data are presented below:

(a) Square-Root Method. The frequency of double-recessive phenotypes in the F_2 may be used as an estimator of the frequency of non-crossover gametes when the F_1 is in coupling phase, and as an estimator of the frequency of crossover gametes when the F_1 is in repulsion phase.

(b) Product-Ratio Method. An estimate of the frequency of recombination from double-heterozygous (dihybrid) F_1 parents can be ascertained from F_2 phenotypes *R-S-*, *R-ss*, *rrS-*, and *rrss* appearing in the frequencies *a*, *b*, *c*, and *d*, respectively. The ratio of crossover to parental types, called the **product ratio**, is a function of recombination.

For coupling data: $x = bc/ad$
For repulsion data: $x = ad/bc$

The recombination fraction represented by the value of x may be read directly from a product-ratio table (Table 6.1). The product-ratio method utilizes all of the F_2 data available and not just the double-recessive class as in the square-root method. The product-ratio method should therefore yield more accurate estimates of recombination than the square-root method.

Table 6.1

Recombination Fraction	Ratio of Products		Recombination Fraction	Ratio of Products	
	ad/*bc* (Repulsion)	*bc*/*ad* (Coupling)		*ad*/*bc* (Repulsion)	*bc*/*ad* (Coupling)
.00	.000000	.000000	.26	.1608	.1467
.01	.000200	.000136	.27	.1758	.1616
.02	.000801	.000552	.28	.1919	.1777
.03	.001804	.001262	.29	.2089	.1948
.04	.003213	.002283	.30	.2271	.2132
.05	.005031	.003629			
.06	.007265	.005318	.31	.2465	.2328
.07	.009921	.007366	.32	.2672	.2538
.08	.01301	.009793	.33	.2892	.2763
.09	.01653	.01262	.34	.3127	.3003
.10	.02051	.01586	.35	.3377	.3259
.11	.02495	.01954	.36	.3643	.3532
.12	.02986	.02369	.37	.3927	.3823
.13	.03527	.02832	.38	.4230	.4135
.14	.04118	.03347	.39	.4553	.4467
.15	.04763	.03915	.40	.4898	.4821
.16	.05462	.04540	.41	.5266	.5199
.17	.06218	.05225	.42	.5660	.5603
.18	.07033	.05973	.43	.6081	.6034
.19	.07911	.06787	.44	.6531	.6494
.20	.08854	.07671	.45	.7013	.6985
.21	.09865	.08628	.46	.7529	.7510
.22	.1095	.09663	.47	.8082	.8071
.23	.1211	.1078	.48	.8676	.8671
.24	.1334	.1198	.49	.9314	.9313
.25	.1467	.1328	.50	1.0000	1.0000

Source: F. R. Immer and M. T. Henderson, "Linkage studies in barley," *Genetics*, 28: 419–440, 1943.

Use of Genetic Maps

Predicting Results of a Dihybrid Cross

If the map distance between any 2 linked genes is known, the expectations from any type of mating may be predicted by use of the gametic checkerboard.

Predicting Results of a Trihybrid Testcross

Map distances or crossover percentages may be treated as any other probability estimates. Given a particular kind of mating, the map distances involved, and either the coincidence or interference for this region of the chromosome, we should be able to predict the results in the offspring generation.

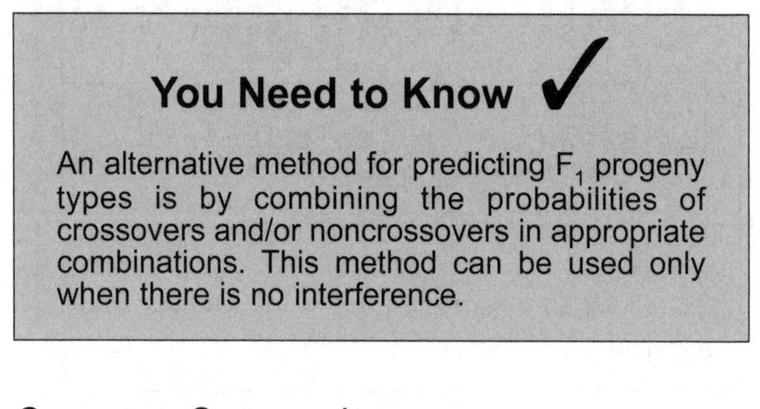

You Need to Know ✓

An alternative method for predicting F_1 progeny types is by combining the probabilities of crossovers and/or noncrossovers in appropriate combinations. This method can be used only when there is no interference.

Crossover Suppression

Many extrinsic and intrinsic factors are known to contribute to the crossover rate. Among these are the effects of sex, age, temperature, proximity to the centromere or **heterochromatic** regions (darkly staining regions presumed to carry little genetic information), chromosomal aberrations such as inversions, and many more. Two specific cases of crossover suppression are presented in this section: (1) complete absence of crossing over in male *Drosophila* and (2) the maintenance of balanced lethal systems as permanent transheterozygotes through the prevention of crossing over.

Absence of Crossing Over in Male *Drosophila*

One of the unusual characteristics of *Drosophila* is the apparent absence of crossing over in males. This fact is shown clearly by the nonequivalent results of reciprocal crosses.

When dihybrid males are crossed to dihybrid females (both in repulsion phase) the progeny will always appear in the ratio 2 : 1 : 1 regardless of the degree of linkage between the genes. The double-recessive class never appears.

Drosophila is not unique in this respect. For example, crossing over is completely suppressed in female silkworms. Other examples of complete and partial suppression of crossing over are common in genetic literature.

Balanced Lethal Systems

A gene that is lethal when homozygous and linked to another lethal with the same mode of action can be maintained in permanent dihybrid condition in repulsion phase when associated with a genetic condition that prevents crossing over. Balanced lethals breed true and their behavior simulate that of a homozygous genotype. These systems are commonly used to maintain laboratory cultures of lethal, semilethal, or sterile mutants.

Balanced lethals may be used to determine on which chromosome an unknown genetic unit resides. Sex-linked genes make themselves known through the nonequivalence of progeny from reciprocal matings. Without the aid of a balanced lethal system, the assignment of an autosomal gene to a particular linkage group may be made through observation of the peculiar genetic ratios obtained from abnormal individuals possessing an extra chromosome (trisomic) bearing the gene under study.

Solved Problem 6.1. In the human pedigree below where the male parent does not appear, it is assumed that he is phenotypically normal. Both hemophilia (*h*) and color blindness (*c*) are sex-linked recessives. Insofar as possible, determine the genotypes for each individual in the pedigree.

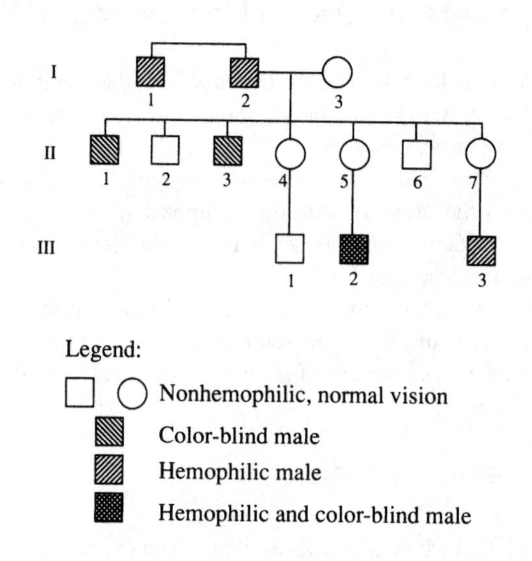

Solution.
Let us begin with males first because, being hemizygous for sex-linked genes, the linkage relationship on their single X chromosome is obvious from their phenotype. Thus, I1, I2, and III3 are all hemophilic with normal color vision and therefore must be *hC/Y*. Nonhemophilic, color-blind males II1 and II3 must be *Hc/Y*. Normal males II2, II6, and III1 must possess both dominant alleles *HC/Y*. III2 is both hemophilic and color blind and therefore must possess both recessives *hc/Y*. Now, let us determine the female genotypes. I3 is normal but produces sons, half of which are color blind and half normal. The X chromosome contributed by I3 to her color-blind sons II1 and II3 must have been *Hc*; the X chromosome she contributed to her normal sons II2 and II6 must have been *HC*. Therefore, the genotype for I3 is *Hc/Hc*.

Normal females II4, II5, and II7 each receive *hC* from their father (I2), but could have received either *Hc* or *HC* on the X chromosome they received from their mother (I3). II4 has a normal son (III1) to which she gives *HC*; therefore II4 is probably *hC*/*HC*, although it is possible for II4 to be *hC*/*Hc* and produce an *HC* gamete by crossing over. II5, however, could not be *hC*/*HC* and produce a son with both hemophilia and color blindness (III2); therefore, II5 must be *hC*/*Hc*, in order to give the crossover gamete *hc* to her son.

Solved Problem 6.2. Two dominant mutants in the first linkage group of the guinea pig govern the traits pollex (*Px*), which is the ativistic return of thumb and little toe, and rough fur (*R*). When dihybrid pollex, rough pigs (with identical linkage relationships) were crossed to normal pigs, their progeny fell into 4 phenotypes: 79 rough, 103 normal, 95 rough, pollex, and 75 pollex. (a) Determine the genotypes of the parents. (b) Calculate the amount of recombination between *Px* and *R*.

Solution.

(a) The parental gametes always appear with greatest frequency, in this case 103 normal and 95 rough, pollex. This means that the 2 normal genes were on one chromosome of the dihybrid parent and the 2 dominant mutations on the other (i.e., coupling linkage).

P:	*Px R*/*px r*	×	*px r*/*px r*
	pollex, rough		*normal*

(b) The 79 rough and 75 pollex types are recombinants, constituting 154 out of 352 individuals = 0.4375 or approximately 43.8% recombination.

Chapter 7
STATISTICAL DISTRIBUTIONS

IN THIS CHAPTER:

- ✔ *The Binomial Expansion*
- ✔ *The Poisson Distribution*
- ✔ *Testing Genetic Ratios*

The Binomial Expansion

In $(p + q)^n$, the p and q represent the probabilities of alternative independent events, and the power n to which the binomial is raised represents the number of trials. Expanding the binomial yields:

$$(p + q)^n = p^2 + 2pq + q^2$$

The sum of the factors in the binomial must add to unity; thus

$$p + q = 1$$

Recall that when two independent events are occurring with the probabilities p and q, then the probability of their joint occurrence is pq. That is, the combined probability is the product of the independent events. When alternative possibilities exist for the satisfaction of the conditions of the problem, the probabilities are combined by addition.

Example 7.1 In two tosses of a coin, with p = heads = ½ and q = tails = ½, there are four possibilities:

First Toss		Second Toss		Probability
Heads (p)	(and)	Heads (p)	=	p^2
Heads (p)	(and)	Tails (q)	=	pq
Tails (q)	(and)	Heads (p)	=	pq
Tails (q)	(and)	Tails (q)	=	q^2
				1.0

which may be expressed as follows:

$$\underset{(2\text{ heads})}{p^2} + \underset{(1\text{ head : }1\text{ tail})}{2pq} + \underset{(2\text{ tails})}{q^2} = 1.0$$

Example 7.2 When a coin is tossed three times, the probabilities for any combination of heads and/or tails can be found from

$$(p + q)^3 = p^3 + 3p^2q + 3pq^2 + q^3$$

Let p = probability of heads = ½ and q = probability of tails = ½.

No. of Heads	No. of Tails	Term	Probability
3	0	p^3	$(\frac{1}{2})^3 = \frac{1}{8}$
2	1	$3p^2q$	$3(\frac{1}{2})^2(\frac{1}{2}) = \frac{3}{8}$
1	2	$3pq^2$	$3(\frac{1}{2})(\frac{1}{2})^2 = \frac{3}{8}$
0	3	q^3	$(\frac{1}{2})^3 = \frac{1}{8}$

The expansion of $(p + q)^3$ is found by manipulating $(p^2 + 2pq + q^2)$ by $(p + q)$. This process can be extended for higher powers, but obviously becomes increasingly laborious. A short method for expanding $(p + q)$ to any power (n) may be performed by following these rules:

(1) The coefficient of the first term is 1. The power of the first factor (p) is n, and that of (q) is 0. (*Note*: Any factor to the zero power is unity.)

(2) Thereafter, in each term, multiply the coefficient by the power of p and divide by the number of that term in the expansion. The result is the coefficient of the *next* term.

(3) Also, thereafter, the power of p will decrease by one and the power of q will increase by one in each term of the expansion.

(4) The fully expanded binomial will have $(n + 1)$ terms. The coefficients are symmetrical about the middle term(s) of the expansion.

Summary:

Term	Coefficient	Powers	
		p	q
1	1	n	0
2	$n(1)/1$	$n - 1$	1
3	$n(n - 1)/1 \cdot 2$	$n - 2$	2
4	$n(n - 1)(n - 2)/1 \cdot 2 \cdot 3$	$n - 3$	3
.	.	.	.
.	.	.	.
.	.	.	.
$n + 1$	1	0	n

Single Terms of the Expansion

The coefficients of the binomial expansion represent the number of ways in which the conditions of each term may be satisfied. The number of combinations (C) of n different things taken k at a time is expressed by

$$ {}_{n}C_{k} = \frac{n!}{(n-k)!k!} \tag{7.1} $$

where $n!$ (called "factorial n") = $n(n-1)(n-2)$ ...1. ($0! = 1$ by definition.)

Formula (*7.1*) can be used for calculating the coefficients in a binomial expansion,

$$ (p+q)^{n} = \sum_{k=0}^{n} {}_{n}C_{k} p^{n-k} q^{k} = \sum_{k=0}^{n} \frac{n!}{(n-k)!k!} p^{n-k} q^{k} $$

where $\sum_{k=0}^{n}$ means to sum what follows as k increases by one unit in each term of the expansion from zero to n. This method is obviously much more laborious than the short method presented previously. However, it does have utility in the calculation of one or a few specific terms of a large binomial expansion. To represent this formula in another way, we can let p = probability of the occurrence of one event (e.g., a success) and q = probability of the occurrence of the alternative event (e.g., a failure); then the probability that in n trials, a success will occur s times and a failure will occur f times is given by

$$ \left(\frac{n!}{s!f!}\right)\left(p^{s}\right)\left(q^{f}\right) $$

The Poisson Distribution

When the probability (p) of a rare event (e.g., a specific mutation) is relatively small and the sample size (n) is relatively large, the binomial distribution is essentially the same as a **Poisson distribution**, but is much easier to solve by the latter.

Another advantage in using the Poisson distribution instead of the binomial distribution is that it allows analysis of data where pn is known, but neither p nor n alone is known. Under these conditions, any term of the binomial expansion is closely approximated by the point Poisson formula

$$\frac{n!}{(n-k)!k!}\,p^k q^{(n-k)} = \frac{e^{-np}(np)^k}{k!}$$

where k is the number of the rare events, q is the probability of the common event (e.g., no mutation), and e is the base of the natural system of logarithms (= 2.71828). The mean (μ) of rare events is equivalent to np. The probabilities that such events happen 0, 1, 2, 3, … times is given by the series

$$\frac{\mu^0 e^{-\mu}}{0!}, \frac{\mu^1 e^{-\mu}}{1!}, \frac{\mu^2 e^{-\mu}}{2!}, \frac{\mu^3 e^{-\mu}}{3!}, \ldots$$

Table 7.1 displays some values of e^{-np} or $e^{-\mu}$ that can be helpful in solving certain problems.

Table 7.1. Values of $e^{-\mu}$
$(0 < \mu < 1)$

μ	0	1	2	3	4	5	6	7	8	9
0.0	1.0000	.9900	.9802	.9704	.9608	.9512	.9418	.9324	.9231	.9139
0.1	.9048	.8958	.8869	.8781	.8694	.8607	.8521	.8437	.8353	.8270
0.2	.8187	.8106	.8025	.7945	.7866	.7788	.7711	.7634	.7558	.7483
0.3	.7408	.7334	.7261	.7189	.7118	.7047	.6977	.6907	.6839	.6771
0.4	.6703	.6636	.6570	.6505	.6440	.6376	.6313	.6250	.6188	.6126
0.5	.6065	.6005	.5945	.5886	.5827	.5770	.5712	.5655	.5599	.5543
0.6	.5488	.5434	.5379	.5326	.5273	.5220	.5169	.5117	.5066	.5016
0.7	.4966	.4916	.4868	.4819	.4771	.4724	.4677	.4630	.4584	.4538
0.8	.4493	.4449	.4404	.4360	.4317	.4274	.4232	.4190	.4148	.4107
0.9	.4066	.4025	.3985	.3946	.3906	.3867	.3829	.3791	.3753	.3716

$(\mu = 1, 2, 3, \ldots, 10)$

μ	1	2	3	4	5	6	7	8	9	10
$e^{-\mu}$	.36788	.13534	.04979	.01832	.006738	.002479	.000912	.000335	.000123	.000045

Note: To obtain values of $e^{-\mu}$ for other values of μ use the laws of exponents.
Example: $e^{-3.48} = (e^{-3.00})(e^{-0.48}) = (.04979)(.6188) = .03081$.

Source: Murray R. Spiegel, *Schaum's Outline of Theory and Problems of Statistics*, McGraw-Hill Book Company, New York, 1961, p. 348.

The variance of the binomial distribution is npq or $np(1-p)$. The Poisson distribution has the same variance, but since p is very small, the probability of the common event $(1-p)$ is almost 1.0. Therefore, in a Poisson distribution, the mean (np) is essentially the same as the variance.

Testing Genetic Ratios

Sampling Theory

If we toss a coin, we expect that half of the time it will land heads up and half of the time tails up. This hypothesized probability is based upon an infinite number of coin tossings wherein the effects of chance deviations from 0.5 in favor of either heads or tails cancel one another. All actual experiments, however, involve finite numbers of observations and therefore some deviation from the expected numbers (sampling error) is to be anticipated. Let us assume that there is no difference between the observed results of a coin-tossing experiment and the expected results that cannot be accounted for by chance alone (**null hypothesis**). How great a deviation from the expected 50-50 ratio in a given experiment should be allowed before the null hypothesis is rejected?

Conventionally, the null hypothesis in most biological experiments is rejected when the deviation is so large that it could be accounted for by chance less than 5 percent of the time. Such results are said to be **significant**. When the null hypothesis is rejected at the 5 percent level, we take 1 chance in 20 of discarding a valid hypothesis. It must be remembered that statistics can never render absolute proof of the hypothesis, but merely sets limits to our uncertainty. If we wish to be even more certain that the rejection of the hypothesis is warranted, we could use the 1 percent level, often called **highly significant**, in which case the experimenter would be taking only one chance in a hundred of rejecting a valid hypothesis.

Sample Size

If our coin-tossing experiment is based on small numbers, we might anticipate relatively large deviations from the expected values to occur quite by chance alone. However, as the sample size increases, the deviation should become proportionately less, so that in a sample of infinite size, the plus and minus chance deviation cancel each other completely to produce the 50-50 ratio.

Degrees of Freedom

Assume a coin is tossed 100 times. We may arbitrarily assign any number of heads from 0 to 100 as appearing in this hypothetical experiment. However, once the number of heads is established, the remainder is tails and must add to 100. In other words, we have $n - 1$ **degrees of freedom** (df) in assigning numbers at random to the n classes within an experiment.

Example 7.3 In an experiment involving three phenotypes ($n = 3$), we can fill two of the classes at random, but the number in the third class must constitute the remainder of the total number of individuals observed. Therefore, we have $3 - 1 = 2$ degrees of freedom.

The number of degrees of freedom in these kinds of problems is the number of variables (n) under consideration minus one. For most genetic problems, the degrees of freedom will be one less than the number of phenotypic classes. Obviously, the more variables involved in an experiment, the greater the total deviation may be by chance.

Chi-Square Test

In order to evaluate a genetic hypothesis, we need a test that can convert deviations from expected values into the probability of such inequalities occurring by chance. Furthermore, this test must also take into consideration the size of the sample and the number of variables (degrees of freedom). The chi-square (χ^2) includes all of these factors.

$$\chi^2 = \sum_{i=1}^{n} \frac{(o_i - e_i)^2}{e_i} = \frac{(o_1 - e_1)^2}{e_1} + \frac{(o_2 - e_2)^2}{e_2} + \cdots + \frac{(o_n - e_n)^2}{e_n}$$

where $\sum_{i=1}^{n}$ means to sum what follows it as the *i* classes increase from 1 to *n*, *o* represents the number of observations within a class, *e* represents the number expected in the class according to the hypothesis under test, and *n* is the number of classes. The value of chi-square may then be converted into the probability that the deviation is due to chance by entering Table 7.2 at the appropriate number of degrees of freedom.

Table 7.2. Chi-Square Distribution

Degrees of Freedom	Probability										
	0.95	0.90	0.80	0.70	0.50	0.30	0.20	0.10	0.05	0.01	0.001
1	0.004	0.02	0.06	0.15	0.46	1.07	1.64	2.71	3.84	6.64	10.83
2	0.10	0.21	0.45	0.71	1.39	2.41	3.22	4.60	5.99	9.21	13.82
3	0.35	0.58	1.01	1.42	2.37	3.66	4.64	6.25	7.82	11.34	16.27
4	0.71	1.06	1.65	2.20	3.36	4.88	5.99	7.78	9.49	13.28	18.47
5	1.14	1.61	2.34	3.00	4.35	6.06	7.29	9.24	11.07	15.09	20.52
6	1.63	2.20	3.07	3.83	5.35	7.23	8.56	10.64	12.59	16.81	22.46
7	2.17	2.83	3.82	4.67	6.35	8.38	9.80	12.02	14.07	18.48	24.32
8	2.73	3.49	4.59	5.53	7.34	9.52	11.03	13.36	15.51	20.09	26.12
9	3.32	4.17	5.38	6.39	8.34	10.66	12.24	14.68	16.92	21.67	27.88
10	3.94	4.86	6.18	7.27	9.34	11.78	13.44	15.99	18.31	23.21	29.59
	Nonsignificant								Significant		

Source: R. A. Fisher and F. Yates, *Statistical Tables for Biological, Agricultural and Medical Research* (6th edition), Table IV, Oliver & Boyd, Ltd., Edinburgh, by permission of the authors and publishers.

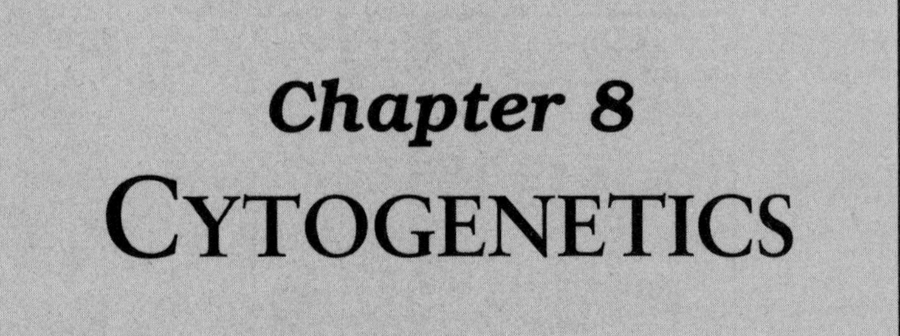

Chapter 8 CYTOGENETICS

IN THIS CHAPTER:

- ✔ *The Union of Cytology with Cytogenetics*
- ✔ *Variation in Chromosome Number*
- ✔ *Variation in the Arrangement of Chromosome Segments*
- ✔ *Variation in the Number of Chromosomal Segments*
- ✔ *Variation in Chromosome Morphology*
- ✔ *Human Cytogenetics*

The Union of Cytology with Cytogenetics

Perhaps one reason Mendel's discoveries were not appreciated by the scientific community of his day was that the mechanics of mitosis and meiosis had not yet been discovered. During the years 1870–1900, rapid advances were made in the study of cells (**cytology**). At the turn of the century, when Mendel's laws were rediscovered, the cytological basis was available to render the statistical laws of genetics intelligible in terms of physical units. **Cytogenetics** is the hybrid science, which attempts to correlate cellular events, especially those of the chromosomes, with genetic phenomena.

Variation in Chromosome Number

Each species has a characteristic number of chromosomes. Most higher organisms are diploid, with two sets of homologous chromosomes: one set donated by the father, the other set by the mother. Variation in the number of sets of chromosomes (**ploidy**) is commonly encountered in nature. It is estimated that one-third of the angiosperms (flowering plants) have more than two sets of chromosomes (polyploid). The term **euploidy** is applied to organisms with chromosomes that are multiples of some basic number (n).

Euploidy

Monoploid. One set of chromosomes (n) is characteristically found in the nuclei of some lower organisms such as fungi. Monoploids in higher organisms are usually smaller and less vigorous than the normal diploids. Few monoploid animals survive. A notable expression exists in male bees and wasps. Monoploid plants are known but are usually sterile.

Triploid. Three sets of chromosomes ($3n$) can originate by the union of a monoploid gamete (n) with a diploid gamete ($2n$). The extra set of chromosomes of the triploid is distributed in various combinations to the germ cells, resulting in genetically unbalanced gametes. Because of the sterility that characterizes triploids, they are not commonly found in natural populations.

Tetraploid. Four sets of chromosomes ($4n$) can arise in body cells by the somatic doubling of the chromosome number. Doubling is accomplished either spontaneously or it can be induced in high frequency by exposure to chemicals such as the alkaloid, colchicine. Tetraploids are also produced by the union of unreduced diploid ($2n$) gametes.

Polyploid. This term can be applied to any cell with more than $2n$ chromosomes. Ploidy levels higher than tetraploid are not commonly encountered in natural populations, but some of our most important crops are polyploid. For example, common bread wheat is hexaploid ($6n$), some strawberries are octaploid ($8n$), etc. Some triploids as well as tetraploids exhibit a more robust phenotype than their diploid counterparts, often having larger leaves, flowers, and fruits (**gigantism**). Many commericial fruits and ornamentals are polyploid. For example, some liver cells of humans are polyploid.

The term **haploid**, strictly applied, refers to the gametic chromosome number. For diploids ($2n$), the haploid number is n. Lower organisms such as bacteria and viruses are called haploids because they have a single set of genetic elements. However, since they do not form gametes comparable to those of higher organisms, the term "monoploid" would seem to be more appropriate.

Aneuploidy

Variations in chromosome number may occur that do not involve whole sets of chromosomes, but only parts of a set. The term **aneuploidy** is given to variations of this nature, and the suffix "-somic" is a part of their nomenclature.

Monosomic. Diploid organisms that are missing one chromosome of a single pair are monosomics with the genomic formula $2n - 1$. The single chromosome without a pairing partner may go to either pole during meiosis, but more frequently will lag at anaphase and fail to be included in either nucleus. Monosomics can thus form two kinds of gametes, (n) and ($n - 1$). In plants, the $n - 1$ gametes seldom function.

Note!

In animals, loss of one whole chromosome often results in genetic unbalance, which is manifested by high mortality or reduced fertility.

Trisomic. Diploids which have one extra chromosome are represented by the chromosomal formula $2n + 1$. One of the pairs of chromosomes has an extra member, so that a trivalent structure may be formed during meiotic prophase. If two chromosomes of the trivalent go to one pole and the third goes to the opposite pole, then gametes will be $(n + 1)$ and (n), respectively. Trisomy can produce different phenotypes, depending upon which chromosome of the complement is present in triplicate. In humans, the presence of one small extra chromosome (autosome 21) has a very deleterious effect resulting in Down syndrome, formerly called "mongolism."

Tetrasomic. When one chromosome of an otherwise diploid organism is present in quadruplicate, this is expressed as $2n + 2$. A quadrivalent may form for this particular chromosome during meiosis.

Double Trisomic. If two different chromosomes are each represented in triplicate, the double trisomic can be symbolized as $2n + 1 + 1$.

Nullosomic. An organism that has lost a chromosome pair is a nullosomic. The result is usually lethal to diploids $(2n - 2)$. Some polyploids, however, can lose two homologues of a set and still survive.

Variation in the Arrangement of Chromosome Segments

Translocations

Chromosomes occasionally undergo spontaneous rupture, or can be induced to rupture in high frequency by ionizing radiation. The broken ends of such chromosomes behave as though they were "sticky" and may rejoin into nonhomologous combinations (**translocations**). A reciprocal translocation involves the exchange of segments between two nonhomologous chromosomes. During meiosis, an individual that is structurally heterozygous for a reciprocal translocation (i.e., two structurally normal chromosomes and two chromosomes that are attached to nonhomologous pieces, as shown in Example 8.1 below, must form a cross-shaped configuration in order to affect pairing or synapsis of all homologous segments. A structural heterozygote may or may not be genetically heterozygous at one or more loci, but this is of no concern for the present purpose. In many of the following diagrams, only chromosomes (not chromatids) are shown and centromeres are omitted for the sake of simplicity.

Example 8.1 Assume that a reciprocal translocation occurs between chromosomes 1-2 and 3-4.

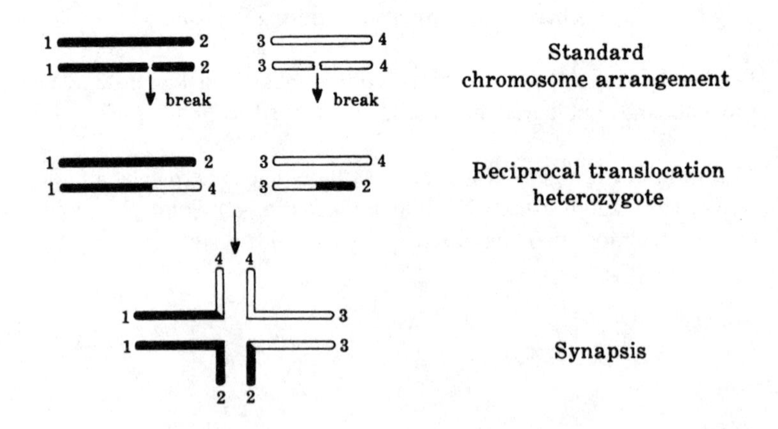

The only way that functional gametes can be formed from a translocation heterozygote is by the alternate disjunction of chromosomes.

Example 8.2 At the end of the meiotic prophase begun in Example 8.1, a ring of four chromosomes is formed. If the adjacent chromosomes move to the poles as indicated in the diagram below, all of the gametes will contain some extra segments (duplications) and some pieces will be missing (deficiencies).

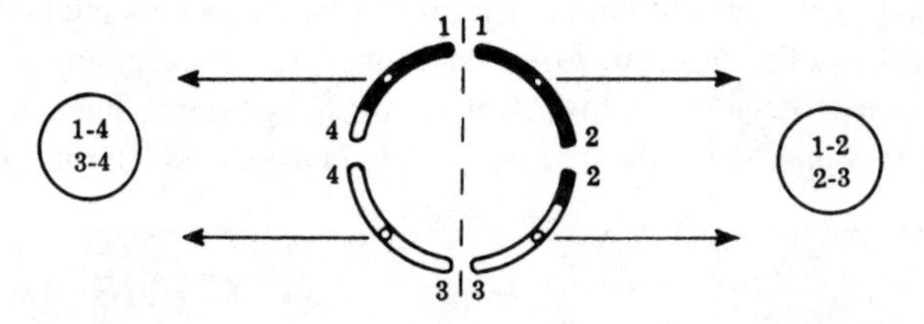

Example 8.3 By forming a "figure-8," alternate disjunction produces functional gametes.

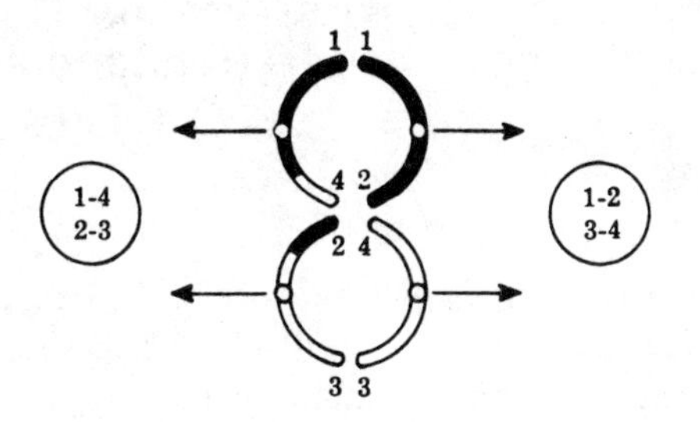

Translocation heterozygotes have several distinctive manifestations: (1) If an organism produces gametes with equal facility by either segregation of adjacent chromosomes (Example 8.2) or by alternate chromosomes (Example 8.3), semisterility occurs because only the latter mechanism produces functional gametes; (2) Some genes that formerly were on nonhomologous chromosomes will no longer appear to be assorting independently; (3) The phenotypic expression of a gene may be modified when it is translocated to a new position in the genome. **Position effects** are particularly evident when genes in euchromatin (lightly staining areas usually containing genetic elements) are shifted near heterochromatic regions (darker staining areas presumably devoid of active genes).

Translocation Complexes. In the evening primrose of the genus *Oenothera*, an unusual series of reciprocal translocations has occurred involving all seven of its chromosome pairs. If we label each chromosome end with a different number, the normal set of seven chromosomes would be 1-2, 3-4, 5-6, 7-8, 9-10, 11-12, and 13-14; a translocation set would be 2-3, 4-5, 6-7, 8-9, 10-11, 12-13, and 14-1. A multiple translocation heterozygote like this would form a ring of 14 chromosomes during meiosis. Different lethals in each of the two haploid sets of seven chromsomes enforces structural heterozygosity. Since only alternate segregation from the ring can form viable gametes, each group of seven chromosomes behaves as though it were a single large linkage group with recombination confined to the pairing ends of each chromosome.

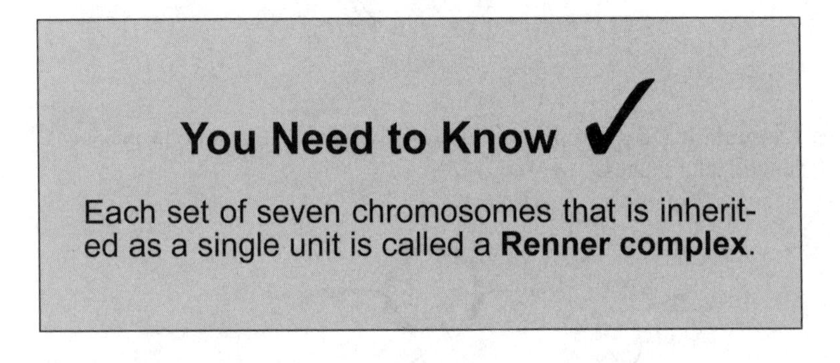
You Need to Know ✓

Each set of seven chromosomes that is inherited as a single unit is called a **Renner complex**.

Example 8.4 In *Oenothera lamarckiana*, one of the Renner complexes is called *gaudens* and the other is called *velans*. This species is largely self-pollinated. The lethals become effective in the zygotic stage so that only the *gaudens-velans* (G-V) zygotes are viable. *Gaudens-gaudens* (G-G) or *velans-velans* (V-V) zygotes are lethal.

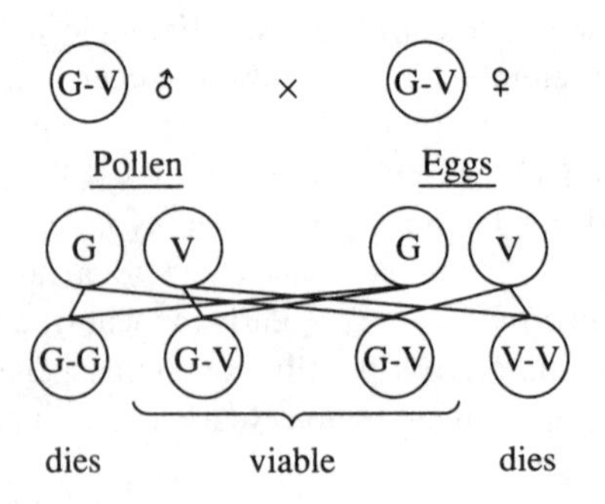

Inversions. Assume that the normal order of segments within a chromosome is (1-2-3-4-5-6) and that breaks occur in regions 2-3 and 5-6, and that the broken piece is reinserted in reverse order. The inverted chromosome now has segments (1-2-3-4-5-6). One way in which inversions might arise is shown in Figure 8-1.

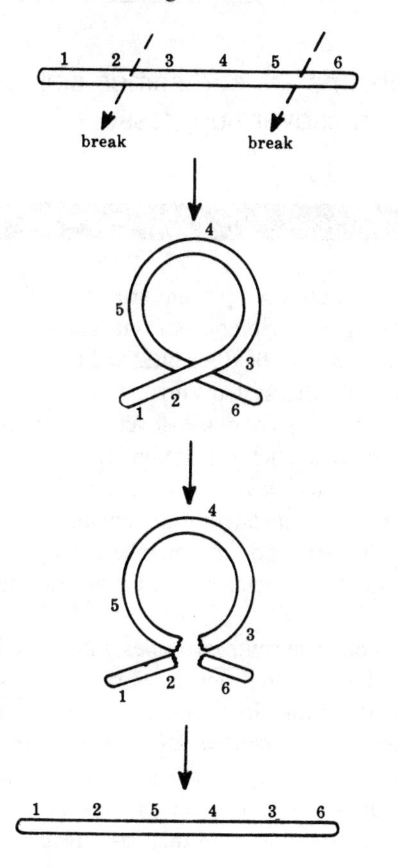

Figure 8-1

An inversion heterozygote has one chromosome in the inverted order and its homologue in the normal order. During meiosis, the synaptic configuration attempts to maximize the pairing between homologous regions in the two chromosomes. This is usually accomplished by a loop in one of the chromosomes. Crossing over within the inverted segment gives rise to crossover gametes which are inviable because of duplications and deficiencies. Chromatids that are not involved in crossing over will be viable. Thus, as we have seen with translocations, inversions produce semisterility and altered linkage relationships.

Inversions are sometimes called "crossover suppressors."

Actually, inversion does not prevent crossovers from occurring but they do prevent the crossover products from functioning. Genes within the inverted segment are thus held together and transmitted as one large linked group. Balanced lethal systems involve either a translocation or an inversion to prevent the recovery of crossover products and thus maintain heterozygosity generation after generation. In some organisms, these "inversions" have a selective advantage under certain environmental conditions and become more prevalent in the population than the standard chromosome order. Two types of inversion heterozygotes will be considered in which crossing over occurs within the inverted segment.

Pericentric Inversion. The centromere lies within the inverted region. First meiotic anaphase figures appear normal unless crossing over occurs within the inversion. If a single two-strand crossover occurs within the inversion, the two chromatids of each chromosome will usually have arms of unequal length (unless there are chromosome segments of equal length on opposite sides of the inversion).

Half of the meiotic products in this case (resulting from crossing over) are expected to contain duplications and deficiencies and would most likely be nonfunctional. The other half of the gametes (noncrossovers) are functional; one-quarter have the normal segmental order, one quarter have the inverted arrangement.

Example 8.5 Assume an inversion heterozygote as shown below with crossing over in region 3-4.

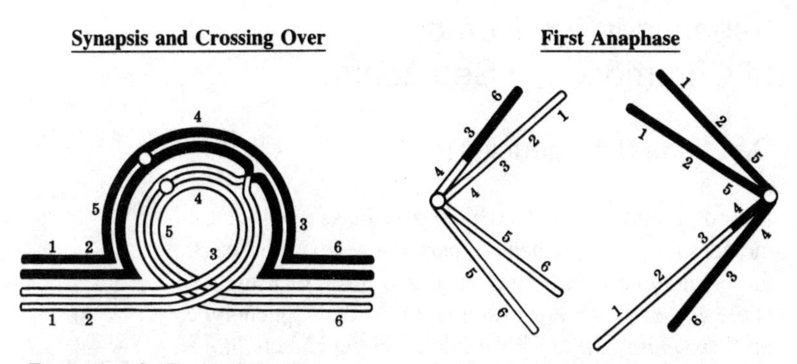

Paracentric Inversion. The centromere lies outside the inverted segment. Crossing over within the inverted segment produces a **dicentric** chromosome (possessing two centromeres) that forms a **bridge** from one pole to the other during first anaphase. The bridge will rupture somewhere along its length and the resulting fragments will contain duplications and/or deficiencies. Also, an **acentric** fragment (without a centromere) will be formed, and since it usually fails to move to either pole, it will not be included in the meiotic products. Again, half of the products are nonfunctional, one-quarter are functional with a normal chromosome, and one-quarter are functional with an inverted chromosome.

Example 8.6 Assume an inversion heterozygote as shown below with crossing over in region 4-5.

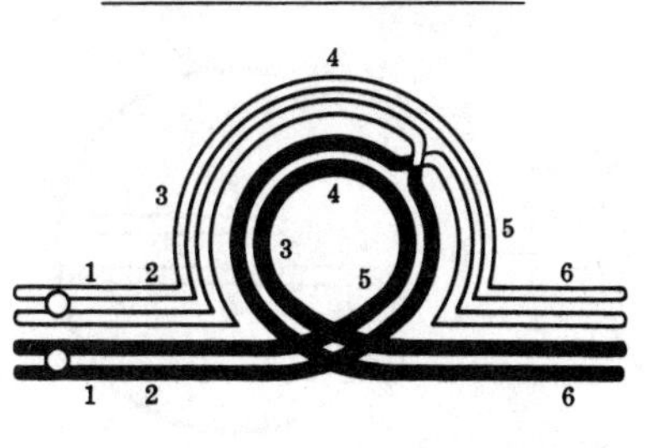

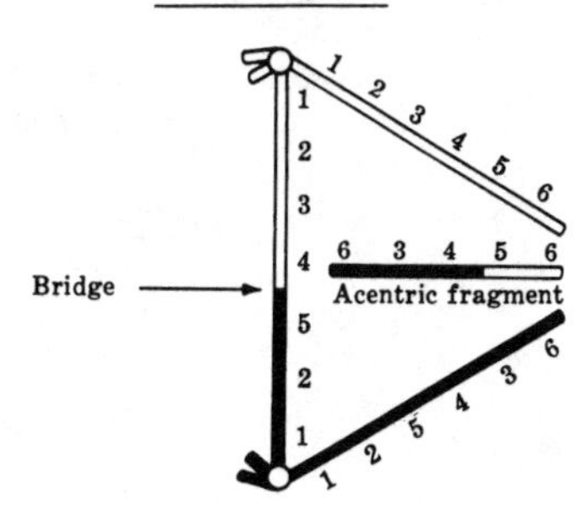

Variation in the Number of Chromosomal Segments

Deletions (Deficiencies)

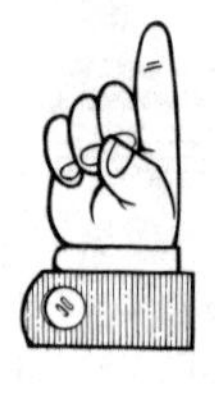

Loss of a chromosomal segment may be so small that it includes only a single gene or part of a gene. In this case, the phenotypic effects may resemble those of a mutant allele at that locus. For example, the "notch" phenotype of *Drosophila* is a sex-linked deletion which acts like a dominant mutation; a deletion at another sex-linked locus behaves as a recessive mutation, producing yellow body color when homozygous. Deletions never backmutate to the normal condition, because a lost piece of chromosome cannot be replaced. In this way, a deletion can be distinguished from a gene mutation. A loss of any considerable portion of a chromosome is usually lethal to a diploid organism because of genetic unbalance. When an organism heterozygous for a pair of alleles, A and a, loses a small portion of the chromosome bearing the dominant allele, the recessive allele on the other chromosome will become expressed phenotypically. This is called **pseudodominance**, but it is a misnomer because the condition is homozygous rather than dizygous at this locus.

Example 8.7 A deficiency in the segment of chromosome bearing the dominant gene A allows the recessive allele to become phenotypically expressed.

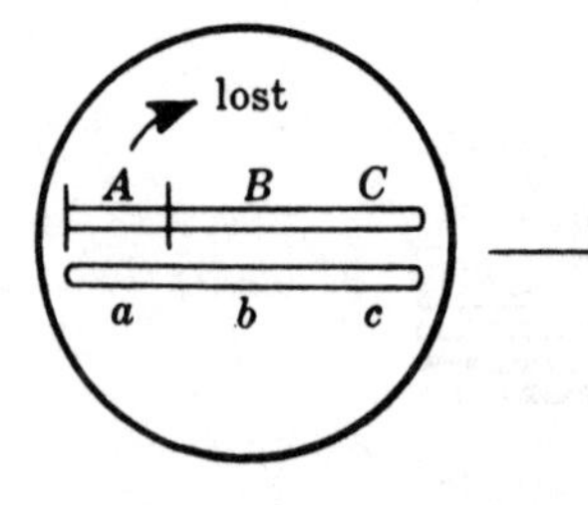

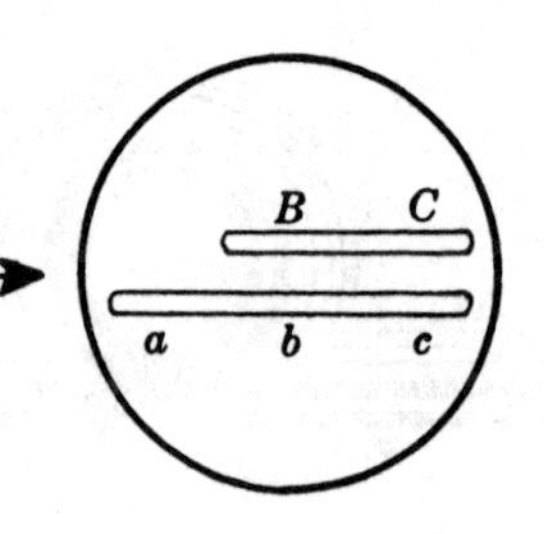

Phenotype:	*ABC* normal heterozygote	*aBC* *a* exhibits pseudodominance

A deletion heterozygote may be detected cytologically during meiotic prophase when the forces of pairing cause the normal chromosome segment to bulge away from the region in which the deletion occurs (Figure 8-2).

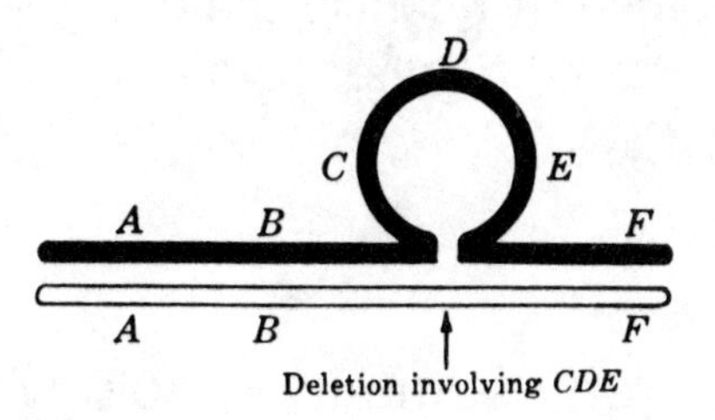

Figure 8-2 Synapsis in a deletion heterozygote

Overlapping deletions have been extensively used to locate the physical position of genes in the chromosome (cytological mapping).

Duplications (Additions)

Extra segments in a chromosome may arise in a variety of ways. Generally speaking, their presence is not as deleterious to the organism as a deficiency. It is assumed that some duplications are useful in the evolution of new genetic material. Because the old genes can continue to provide for the present requirements of the organism, the superfluous genes may be free to mutate to new forms without a loss in immediate adaptability. Genetic redundancy, of which this is one type, may protect the organism from the effects of a deleterious recessive gene or from an otherwise lethal deletion. During meiotic pairing, the chromosome bearing the duplicated segment bulges away from its normal homologue to maximize the juxtaposition of homologous regions. In some cases, extra genetic material is known to cause a distinct phenotypic effect.

Remember

Relocation of chromosomal material without altering in quantity may result in an altered phenotype (position effect).

Variation in Chromosome Morphology

Isochromosomes

It has already been shown that a translocation can change the structure of the chromosome both genetically and morphologically. The length of the chromosome may be longer or shorter, depending upon the size of the translocated piece. An inversion does not normally change the length of the chromosome, but if the inversion includes the centromere (pericentric), the position of the centromere may be changed considerably. Deletions or duplications, if viable, may sometimes be detected cytologically by a change in the size of the chromosome (or banding pattern in the case of the giant chromosomes of *Drosophila*), or by the presence of "bulges" in the pairing figure. Chromsomes with unequal arm lengths may be changed to **isochromosomes** having arms of equal length and genetically homologous with each other, by an abnormal transverse division of the centromere. The telocentric X chromosome of *Drosophila* may be changed to an "attached-X" form by a misdivision of the centromere (Figure 8-3).

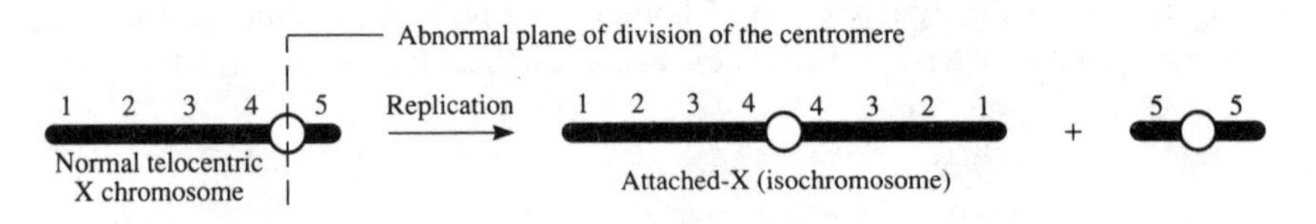

Figure 8-3 Origin of attached-X chromosome. Segment 5 is nonessential heterochromatin.

Bridge-Breakage-Fusion-Bridge Cycles

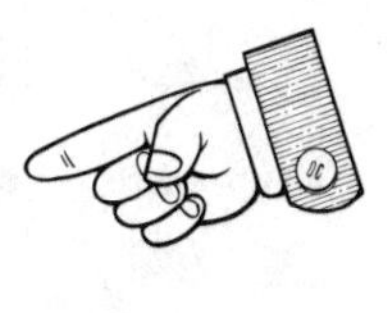

The shape of a chromosome may change at each division once it has broken. Following replication of a broken chromosome, the broken ends of the sister chromatids may be fused by DNA repair mechanisms. Such broken ends are said to be "sticky." When the chromatids move to opposite poles, a bridge is formed. The bridge will break somewhere along its length and the cycle repeats at the next division. This sequence of events is called the **bridge-breakage-fusion-bridge cycle**. Mosaic tissue appearing as irregular patches of an unexpected phenotype on a background of normal tissue (**variegation**) can be produced by such a cycle. The size of the unusual tissue generally bears an inverse relationship to the period of development at which the original break occurred; i.e., the earlier the break occurs, the larger will be the size of the abnormal tissue.

Ring Chromosomes

Chromosomes are not always rod-shaped. Occasionally, ring chromosomes are encountered in plants or animals. If breaks occur at each end of a chromosome, the broken ends may become joined to form a ring chromosome (Figure 8-4). If an acentric fragment is formed by union of the end pieces, it will soon be lost. The phenotypic consequences of these deletions vary, depending on the specific genes involved. Crossing over between ring chromosomes can lead to bizarre anaphase figures.

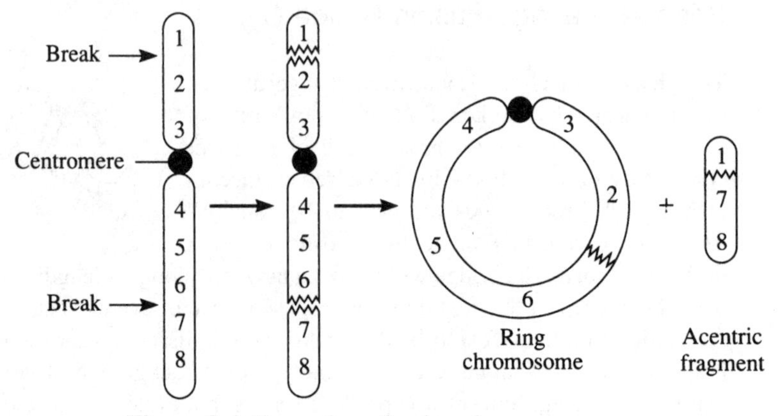

Figure 8-4. Formation of a ring chromosome

Robertsonian Translocation

A whole arm fusion (**Robertsonian translocation**) is a eucentric, reciprocal translocation between two acrocentric chromosomes where the break in one chromosome is near the front of the centromere and the break in the other chromosome is immediately behind its centromere. The smaller chromosome thus formed consists of largely inert heterochromatic material near the centromeres; it usually carries no essential genes and tends to become lost. A Robertsonian translocation thus results in a reduction of the chromosome number (Figure 8-5).

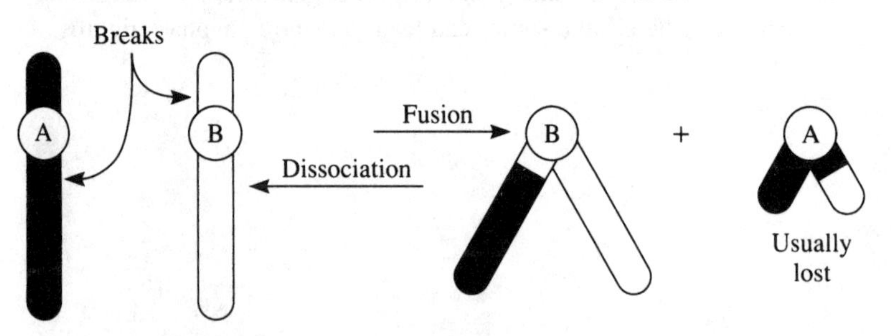

Figure 8-5. Formation of a metacentric chromosome by fusion of two acrocentric chromosomes (Robertsonian translocation). Dissociation is not possible if the small chromsome is lost

Human Cytogenetics

The diploid human chromosome number of 46 (23 pairs) was established by Tijo and Levan in 1956. When grouped as homologous pairs, the somatic chromosome complement (**karyotype**) of a cell becomes an **idiogram**. Formerly, a chromosome could be distinguished only by its length and the position of its centromere at the time of maximum condensation (late prophase). No single autosome could be easily identified, but a chromosome could be assigned to one of seven groups (A-G). Group A consists of large, metacentric chromosomes (1-3); group B contains submedian chromosomes (4-5); group C has medium-sized chromosomes with submedian centromeres (pairs 6-12); group D consists of medium-sized chromosomes (pairs 13-15) with one very short arm (acrocentric); chromosomes in group E (16-18) are a little shorter than in group D with median or submedian centromeres; group F (19-20) contains short, metacentric chromosomes; and Group G has the smallest acrocentric chromosomes (21,22). The X and Y sex chromosomes are not members of the autosome groups, and are usually placed together in one part of the idiogram. The Y chromosome may vary in size from one individual to another but usually has the appearance of G-group autosomes. The X chromosome has the appearance of a group C autosome.

More recently, special staining techniques (e.g., Giemsa, quinacrine) have revealed specific banding patterns (G bands, Q bands, etc.) for each chromosome, allowing individual identification of each chromosome in the karyotype (Figure 8-6).

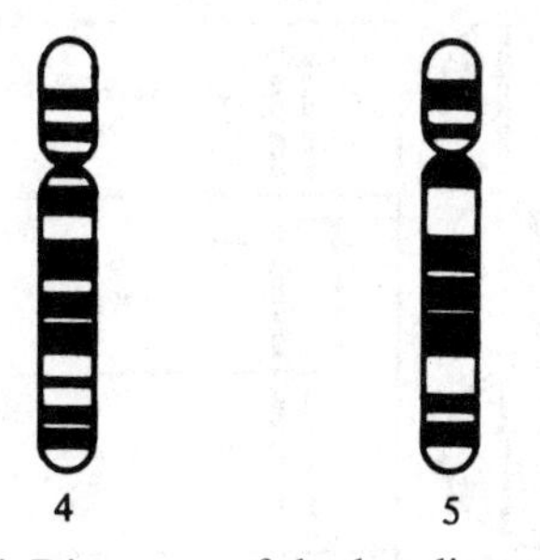

Figure 8-6. Diagrams of the banding patterns that distinguish human chromosomes 4 and 5

Solved Problem 8.1. In 1931, Stern found two different translocations in *Drosophila* from which he developed females possessing heteromorphic X chromosomes. One X chromosome had a piece of the Y chromosome attached to it; the other X was shorter and had a piece of chromosome IV attached to it. Two sex-linked genes were used as markers for detecting crossovers, the recessive trait carnation eye color (*car*) and the dominant trait bar eye (*B*). Dihybrid bar females with heteromorphic chromosomes (both mutant alleles on the X portion of the X-IV chromosome) were crossed to homozygous carnation males with normal chromosomes. The results of this experiment provided cytological proof that genetic crossing over involves an actual physical exchange between homologous chromosome segments. Diagram the expected cytogenetic results of this cross showing all genotypes and phenotypes.

Solution:

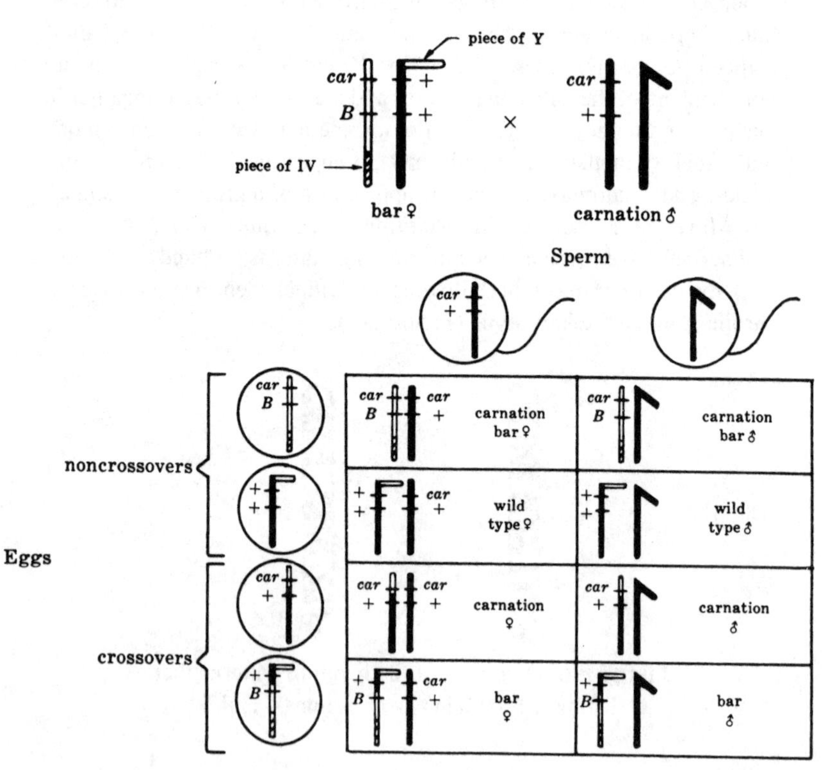

The existence of a morphologically normal X chromosome in recombinant male progeny with carnation eyes provides cytological proof that genetic crossing over is correlated with physical exchange between homologous chromosomes in the parents. Similarly, all other phenotypes correlate with the cytological picture. Chromosomal patterns other than the ones shown above may be produced by crossing over outside the inverted region.

Chapter 9
Quantitative Genetics and Breeding Principles

In This Chapter:

- ✔ *Qualitative vs. Quantitative Traits*
- ✔ *The Normal Distribution*
- ✔ *Types of Gene Action*
- ✔ *Heritability*
- ✔ *Selection Methods*
- ✔ *Mating Methods*

Qualitative vs. Quantitative Traits

The classical Mendelian traits encountered in the previous chapters have been qualitative in nature, i.e., traits that are easily classified into distinct phenotypic categories. These discrete phenotypes are under the genetic control of only one or a very few genes with little or no environmental modification to obscure the gene effects. In contrast to this, the variability exhibited by many agriculturally important traits fails to fit into separate phenotypic classes (discontinuous variability), but instead forms a spectrum of phenotypes that blend imperceptively from one type to another (continuous variability). Economically important traits such as body weight gains, mature plant heights, egg or milk production records, and yield of grain per acre are **quantitative**, or **metric**, traits with continuous variability.

The basic difference between qualitative and quantitative traits involves the number of genes contributing to the phenotypic variability and the degree to which the phenotype can be modified by environmental factors. Quantitative traits may be governed by many genes (perhaps 10 – 100 or more), each contributing such a small amount to the phenotype that their individual effects cannot be detected by Mendelian methods. Genes of this nature are called **polygenes**. All genes act in concert with other genes. Thus, more than one gene may contribute to a given trait. Furthermore, each gene usually has effects on more than one trait (pleiotropy). The idea that each character is controlled by a single gene (the one-gene-one-trait hypothesis) has often been falsely attributed to Mendel. For example, he observed that purple flowers are correlated with brown seeds and a dark spot on the axils of leaves; similarly, white flowers are correlated with light-colored seeds and no axillary spots on the leaves.

For a given gene, some of its pleiotropic effects may be relatively strong for certain traits, whereas its effects on other traits may be so weak that they are difficult or impossible to identify by Mendelian techniques. It is the totality of these pleiotropic effects of numerous loci (polygenes) that constitutes the genetic base of a quantitative character. In addition to this genetic component, the phenotypic variability of a quantitative trait in a population usually has an environmental component. It is the task of the geneticist to determine the magnitude of the genetic and environmental components of the total phenotypic variability of each quantitative trait in a population. In order to accomplish this task, use is made of some rather sophisticated mathematics, especially of statistics. Below are summarized some of the major differences between qualitative and quantitative genetics.

Qualitative Genetics	Quantitative Genetics
1. Characters of *kind.*	1. Characters of *degree.*
2. *Discontinuous* variation; discrete phenotypic classes	2. *Continuous* variation; phenotypic measurements form a spectrum
3. *Single-gene* effects discernible	3. *Polygenic* control; effects of single genes too slight to be detected
4. Concerned with *individual matings* and their progeny	4. Concerned with a *population* of organisms consisting of all possible kinds of matings
5. Analyzed by making *counts and ratios*	5. Statistical analyses give estimates of *population parameters* such as the mean and standard deviation

The Normal Distribution

The study of a quantitative trait in a large population usually reveals that very few individuals possess the extreme phenotypes and that progressively more individuals are found nearer the average value for that population. This type of symmetrical distribution is characteristically bell-shaped as shown in Figure 9-1 and is called a **normal distribution**. It is approximated by the binomial distribution $(p + q)^n$ when the power of the binomial is very large and p and q are both $1/n$ or greater.

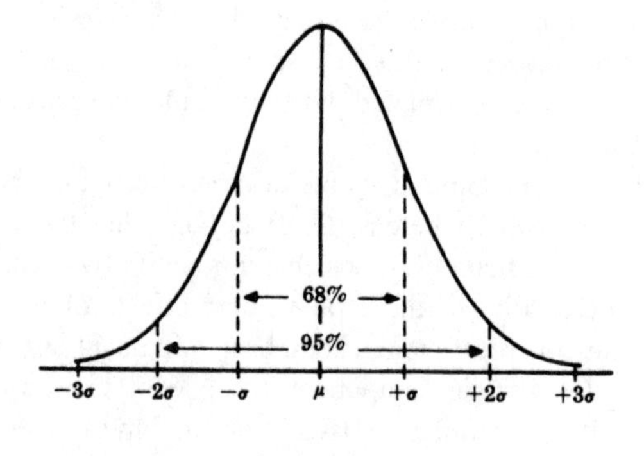

Figure 9-1

Average Measurements

The average phenotypic value for a normally distributed trait is expressed as the **arithmetic mean** ($\overline{X}$, read "X bar"). The arithmetic mean is the sum of the individual measurements (ΣX) divided by the number of individuals measured (N). The Greek letter "sigma" (Σ) directs the statistician to sum what follows.

$$\overline{X} = \frac{\sum_{i=1}^{N} X_i}{N} = \frac{X_1 + X_2 + X_3 + \cdots + X_N}{N}$$

It is usually not feasible to measure every individual in a population; therefore, measurements are usually made on a sample from that population in order to estimate the population value (parameter). If the sample is truly representative of the larger population of which it is a part, then $\overline{X}$ will be an accurate estimate of the mean of the entire population (μ).

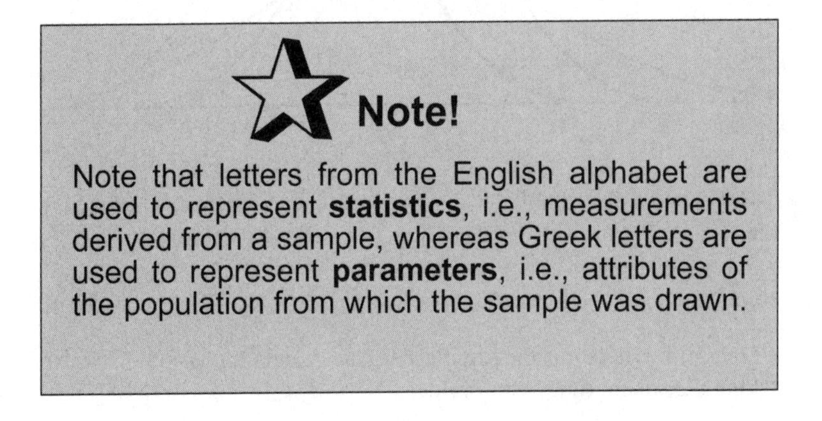

Note!

Note that letters from the English alphabet are used to represent **statistics**, i.e., measurements derived from a sample, whereas Greek letters are used to represent **parameters**, i.e., attributes of the population from which the sample was drawn.

Parameters are seldom known and must be estimated from results gained by sampling. Obviously, the larger the sample size, the more accurately the statistic estimates the parameter.

Measurement of Variability

Consider the three normally distributed populations shown in Figure 9-2. Populations A and C have the same mean, but C is much more variable than A. A and B have different means, but otherwise appear to have the same shape (dispersion). Therefore, in order to adequately define a normal distribution, we must know not only its mean but also how much variability exists. One of the most useful measures of variability in a population for genetic purposes is the **standard deviation**, symbolized by the lowercase Greek letter "sigma" (σ). A sample drawn from this population at random will have a sample standard deviation (s).

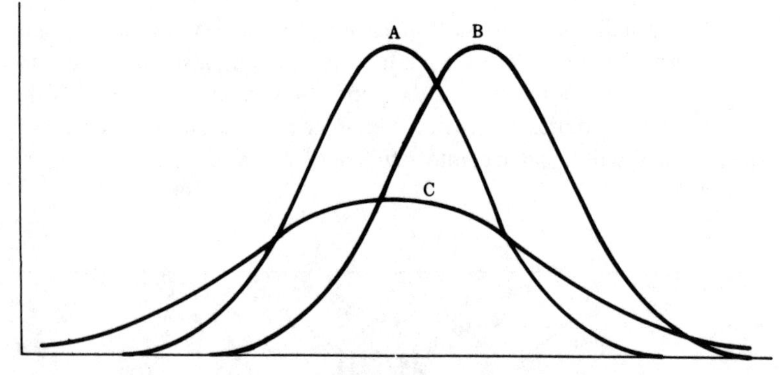

Figure 9-2

To calculate s, the sample mean ($\overline{X}$) is subtracted from each individual measurement (X_i) and the deviation ($X_1 - \overline{X}$) is squared $(X_1 - \overline{X})^2$, summed over all individuals in the sample $\sum_{i=1}^{n}\left[\left(X_i - \overline{X}\right)^2\right]$, and divided by $n - 1$, where n is the sample size.The calculation is completed by taking the square root of this value.

$$s = \sqrt{\frac{\sum_{i=1}^{n}\left(X_i - \overline{X}\right)^2}{n-1}}$$

To calculate σ, we substitute the total population size (N) for n in the above formula. For samples less than about 30, the appropriate correction factor for the denominator should be $n - 1$; for sample sizes greater than this, it makes little difference in the value of s whether n or $n - 1$ is used in the denominator. All other things being equal, the larger the sample size, the more accurately the statistic s should estimate the parameter σ.

Coefficient of Variation. Traits with relatively large average metric values generally are expected to have correspondingly larger standard deviations than traits with relatively small average metric values. Furthermore, since different traits may be measured in different units, the **coefficient of variation** is useful for comparing their relative variabilities. Dividing the standard deviation by the mean renders the coefficient of variation independent of the units of measurement.

$$\text{Coefficient of variation} = \sigma/\mu \text{ for a population}$$
$$= s/\overline{X} \text{ for a sample}$$

Variance

The square of the standard deviation is called **variance** (σ^2). Unlike the standard deviation, however, variance cannot be plotted on the normal curve and can only be represented mathematically. Variance is widely used as an expression of variability because of the additive nature of its components. By a technique called "analysis of variance," the total phenotypic variance (σ^2_P) expressed by a given trait in a population can be statistically fragmented or partitioned into components of genetic variance (σ^2_G) nongenetic (or environmental) variance (σ^2_E), and variance due to genotype-environment interactions (σ^2_{GE}). Thus,

$$\sigma^2_P = \sigma^2_G + \sigma^2_E = \sigma^2_{GE}$$

Variance Method of Estimating the Number of Genes. A population such as a line, a breed, a variety, a strain, a subspecies, etc., is composed of individuals that are more nearly alike in their genetic composition than those in the species as a whole. Phenotypic variability will usually be expressed even in a group of organisms that are genetically identical. All such variability within pure lines is obviously environmental in origin. Crosses between two pure lines produce a genetically uniform hybrid F_1. Phenotypic variability in the F_1 is likewise nongenetic in origin. In the formation of the F_2 generation, gene combinations are reshuffled and dealt out in new combinations to the F_2 individuals. It is a common observation that the F_2 generation is much more variable than the F_1 from which it was derived.

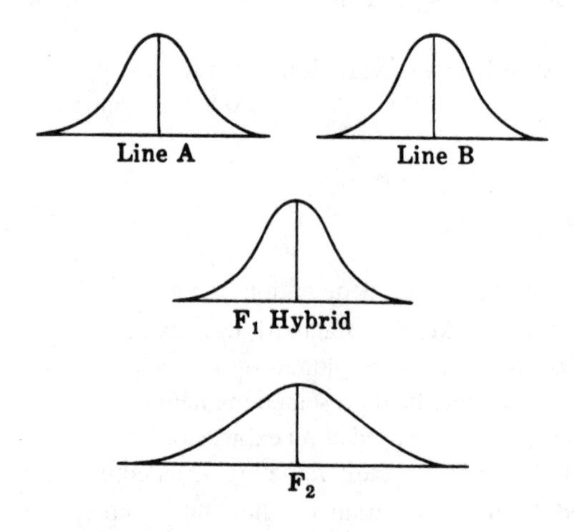

In a normally distributed trait, the means of the F_1 and F_2 populations tend to be intermediate between the means of the two parental lines. If there is no change in the environment from one generation to the next, then the environmental variation of F_2 should be approximately the same as that of the F_1. An increase in phenotypic variance of the F_2 over that of the F_1 may then be attributed to genetic causes. Thus, the genotypic variance of the F_2 (σ^2_{GF2}) is equal to the phenotypic variance of the F_2 (σ^2_{PF2}) minus the phenotypic variance of the F_1 (σ^2_{PF1}):

$$\sigma^2_{GF2} = \sigma^2_{PF2} - \sigma^2_{PF1}$$

The genetic variance of the F_2 is expressed by the formula $\sigma^2_{GF2} = (a^2N)/2$, where a is the contribution of each active allele and N is the number of *pairs* of genes involved in the metric trait. An estimate of a is obtained from the formula $a = D/2N$, where D is the numerical difference between the two parental means. Making substitutions and solving for N,

$$\sigma^2_{PF2} - \sigma^2_{PF1} = \sigma^2_{GF2} = a^2N/2 = D^2/8N$$

from which

$$N = \frac{D^2}{8\left(\sigma^2_{PF2} - \sigma^2_{PF1}\right)}$$

This formula is an obvious oversimplification since it assumes all genes are contributing cumulatively the same amount to the phenotype, no dominance, no linkage, and no interaction.

Types of Gene Action

Alleles may interact with one another in a number of ways to produce variability in their phenotypic expression. The following models may help in understanding various models of gene action.

(1) With dominant lacking, i.e., **additive genes**, each A^1 allele is assumed to contribute nothing to the phenotype (null allele), whereas each A^2 allele contributes one unit to the phenotype (active allele).

Scale of phenotypic value:	0	1	2
Genotype:	A^1A^1	A^1A^2	A^2A^2

(2) With **partial or incomplete dominance**, the heterozygote is almost as valuable as the A^2A^2 homozygote.

Scale of phenotypic value:	0	1.5	2
Genotype:	A^1A^1	A^1A^2	A^2A^2

(3) In **complete dominance**, identical phenotypes are produced by the heterozygote and A^2A^2 homozygote.

Scale of phenotypic value:	0	2
Genotype:	A^1A^1	A^1A^2 A^2A^2

(4) In **overdominance**, the heterozygote is more valuable than either homozygous genotype.

Scale of phenotypic value:	0	2	2.5
Genotype:	A^1A^1	A^2A^2	A^1A^2

If allelic interaction is completely additive, a linear phenotypic effect is produced. In Figure 9-3, a constant increment (i) is added to the phenotype for each A^2 allele in the genotype.

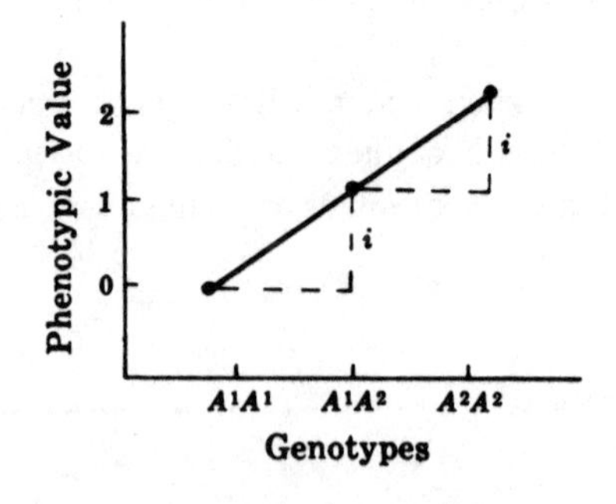

Figure 9-3

Even if complete dominance is operative, an underlying component of additivity (linearity) is still present (solid line in Figure 9-4). The derivations from the additive scheme (dotted lines) due to many such genes with partial or complete dominance can be statistically estimated from appropriately designed experiments. The genetic contributions from such effects appear in the dominance component of variance (σ^2_D).

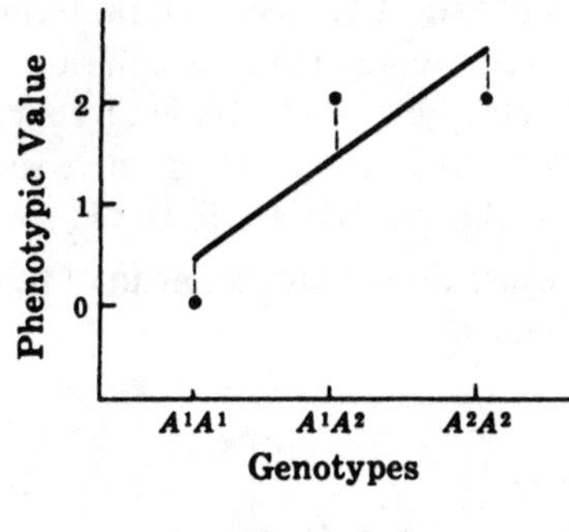

Figure 9-4

In a much more complicated way, deviations from an underlying additive scheme could be shown to exist for the interactions between genes at different loci (epistatic relationships). The contribution to the total genetic variance (σ^2_G) made by these genetic elements can be partitioned into a component called the epistatic or interaction variance (σ^2_I).

The sum of the additive gene effects produced by genes lacking dominance (additive genes) and by the additive contribution of genes with dominance or epistatic effects appears in the additive component of genetic variance (σ^2_A).

Thus, the total genetic variance can be partitioned into three fractions:

$$\sigma^2_G = \sigma^2_A + \sigma^2_D + \sigma^2_I$$

Additive vs. Multiplicative Gene Action

Additive gene action produces an arithmetic series of phenotypic values such as 2, 4, 6, 8, … representing the contribution of 1, 2, 3, 4, … active alleles, respectively. Additive gene action tends to produce a normal phenotype distribution with the mean of the F_1 intermediate between means of the two parental populations. However, not all genes act additively. Some exhibit **multiplicative gene action**, forming a geometric series such as 2, 4, 8, 16, … representing the contributions of 1, 2, 3, 4, … active alleles, respectively. Traits governed by multiplicative gene action tend to be skewed into an asymmetrical curve such as that shown for the F_2 in Figure 9-5. The means of the F_1 and F_2 are nearer to one of the parental means because the **geometric mean** of two numbers is the square root of their product.

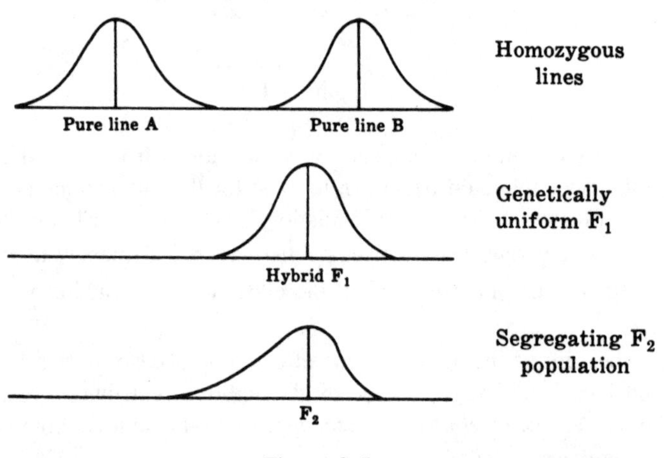

Figure 9-5

If a skewed distribution can be converted to a normal distribution by merely transforming the data to a logarithmic scale, this is evidence for multiplicative gene action.

The variance and the mean are independent parameters in a normal distribution. That is, if the population mean is increased, we cannot predict in advance to what degree the variance will be increased. In the case of multiplicative gene action, however, the variance is dependent upon the mean so that as the mean increases, the variance increases proportionately. The coefficients of variation in segregating populations thereby remain constant.

The concepts of heritability and selection theory discussed in the following sections will deal only with normal distributions.

Heritability

One of the most important factors in the formulation of effective breeding plans for improving the genetic quality of crops and livestock is a knowledge of the relative contribution made by genes to the variability of a trait under consideration. The variability of phenotypic values for a quantitative trait can, at least in theory, be partitioned into genetic and nongenetic (environmental) components.

$$\sigma_P{}^2 = \sigma_G{}^2 + \sigma_E{}^2$$

Heritability (symbolized h^2 or H in some texts) is the proportion of the total phenotypic variance due to gene effects.

$$h^2 = \frac{\sigma_G^2}{\sigma_P^2}$$

The heritability of a given trait may be any number from 0 to 1.

The parameter of heritability involves all types of gene action and thus forms a broad estimate of heritability. In the case of complete dominance, when a gamete bearing the active dominant allele A^2 unites with a gamete bearing the null allele A^1, the resulting phenotype might be two units. When two A^2 gametes unite, the phenotypic result would still be two units. On the other hand, if genes lacking dominance (additive genes) are involved, then the A^2 gamete will add one unit to the phenotype of the resulting zygote, regardless of the allelic contribution of the gamete with which it unites. Thus, only the additive genetic component of variance has the quality of predictability necessary in the formulation of breeding plans. Heritability, in this narrower sense, is the ratio of the additive genetic variance to the phenotypic variance:

$$h^2 = \frac{\sigma_A^2}{\sigma_P^2}$$

It must be emphasized that the heritability of a trait applies only to a given population living in a particular environment. A genetically different population (perhaps a different variety, breed, race, or subspecies of the same species) living in an identical environment is likely to have a different heritability for the same trait. Likewise, the same population is likely to exhibit different heritabilities for the same trait when measured in different environments because a given phenotype does not always respond to different environments in the same way. There is no one genotype that is adaptively superior in all possible environments. That is why natural selection tends to create genetically different populations within a species, each population being specifically adapted to local conditions rather than generally adapted to all environments in which the species is found.

Selection Methods

Artificial selection is operative when humans determine which individuals will be allowed to leave offspring (and/or the number of such offspring). Likewise, **natural selection** allows only those individuals to reproduce that possess traits adaptive to the environments in which they live. There are several methods by which artificial selection can be practiced.

Mass Selection

If heritability of a trait is high, most of the phenotypic variability is due to genetic variation. Thus, a breeder should be able to make good progress by selecting from the masses those that excel phenotypically because the offspring-parent correlation should be high. This is called **mass selection**, but it is actually based on the individual's own performance record or phenotype. As the heritability of a trait declines, so does the prospect of making progress in improving the genetic quality of the selected line. In practice, selection is seldom made on the basis of one characteristic alone. Breeders usually desire to practice selection based on several criteria simultaneously. However, the more traits selected for, the less selection "pressure" can be exerted on each trait. Selection should thus be limited to the two or three traits which the breeder considers to be the most important economically. It is probable that individuals scoring high in trait A will be mediocre or even poor in trait B (unless the two traits have a positive genetic correlation, i.e., some of the genes increasing trait A are also contributing positively to trait B).

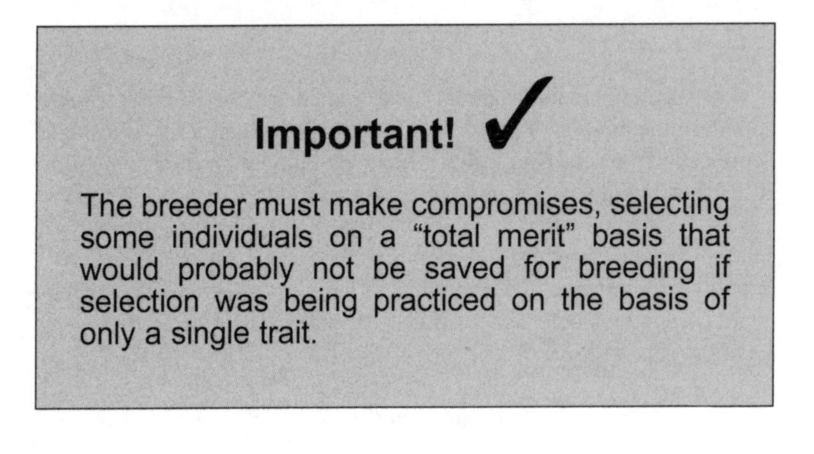

The model used to illustrate the concept of genetic gain, wherein only individuals that score above a certain minimum value for a single trait would be saved for breeding, must now be modified to represent the more probable situation in which selection is based on the total merit of two or more traits (Figure 9-6).

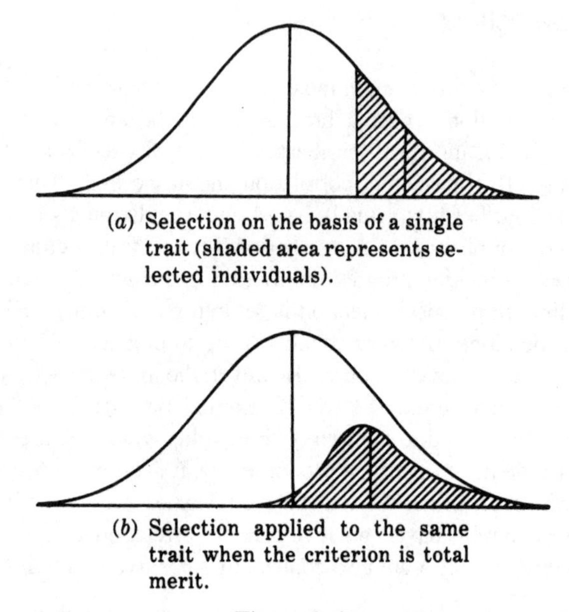

(*a*) **Selection on the basis of a single trait (shaded area represents selected individuals).**

(*b*) **Selection applied to the same trait when the criterion is total merit.**

Figure 9-6

In selecting breeding animals on a "total merit" basis, it is desirable to reduce the records of performance on the important traits to a single score called the **selection index**. The index number has no meaning by itself, but is valuable in comparing several individuals on a *relative* basis. The methods used in constructing an index may be quite diverse, but they usually take into consideration, the heritability and the relative economic importance of each trait in addition to the genetic and phenotypic correlations between the traits. An index (I) for three traits may have the general form

$$I = aA' + bB' + cC'$$

Where a, b, and c are coefficients correcting for the relative heritability and the relative economic importance for traits A, B, and C, respectively, and where A′, B′, and C′ are the numerical values of traits A, B, and C expressed in "standardized form." A **standardized variable** (X′) is computed in a sample by the formula

$$X' = \frac{X - \overline{X}}{s}$$

where X is the record of performance made by an individual, $\overline{X}$ is the average performance of the population, and s is the standard deviation of the trait. In comparing different traits, one is confronted by the fact that the mean and variability of each trait is different and often the traits are not even expressed in the same units.

The standardized variable, however, is a pure number (i.e., independent of the units used) based on the mean and standard deviation. Therefore, any production record or score of a quantitative nature can be added to any other such trait if they are expressed in standardized form.

Family Selection

When both broad and narrow heritabilities of a trait are low, environmental variance is high compared to genetic variance. Family selection is most useful when heritabilities of traits are low and family members resemble one another only because of their genetic relationship. It is usually more practical to first reduce environmental variance by changing the farming or husbandry practices before initiating selective breeding programs. Another way to minimize the effects of an inflated environmental variance is to save for breeding purposes all members of families that have the highest average performance even though some members of such families have relatively poor phenotypes. In practice, it is not uncommon to jointly use more than one selection method, e.g., choosing only the top 50 percent of individuals in only the families with the highest averages.

Family selection is most beneficial when members of a family have a high average genetic relationship to one another but the observed resemblance is low. If inbreeding increases the average genetic relationship within a family more than the increases in phenotypic resemblance, the gain from giving at least some weight to family averages may become relatively large.

Pedigree Selection

In this method, consideration is given to the merits of ancestors. Rarely should pedigree selection be given as much weight as the individual's own merit unless the selected traits have low inheritabilities and the merits of the parents and grandparents are much better known than those of the individual in question.

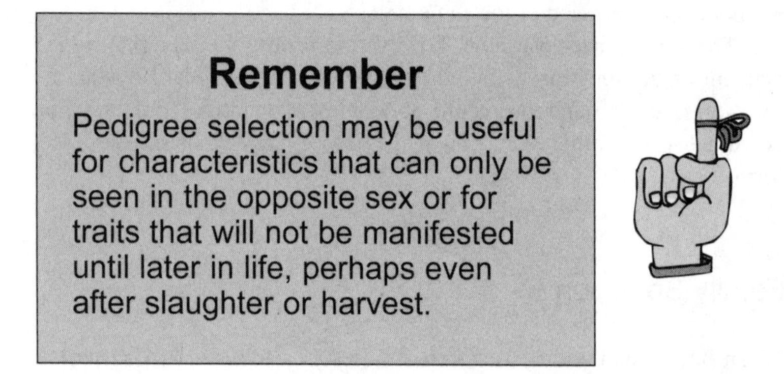

The value of pedigree selection depends upon how closely related the ancestor is to the individual in the pedigree, upon how many ancestors' or colateral ancestors' records exist, upon how completely the merits of such ancestors are known, and upon the degree of heritability of the selected traits.

Progeny Test

A **progeny test** is a method of estimating the breeding value of an animal by the performance or phenotype of its offspring. It has its greatest utility for those traits that (1) can be expressed only in one sex (e.g., estimating the genes for milk production possessed by a bull), (2) cannot be measured until after slaughter (e.g., carcass characteristics), or (3) have low heritabilities so that individual selection is apt to be highly inaccurate.

Progeny testing cannot be practiced until after the animal reaches sexual maturity. In order to progeny-test a male, he must be mated to

several females. If the sex ratio is 1 : 1, then obviously every male in a flock or herd cannot be tested. Therefore, males that have been saved for a progeny test have already been selected by some other criteria earlier in life. The more progeny each male is allowed to produce the more accurate the estimate of his "transmitting ability" (**breeding value**), but in so doing, fewer males can be progeny-tested. If more animals could be tested, the breeder would be able to save only the very best for widespread use in the herd or flock. Thus, a compromise must be made, in that the breeder fails to test as many animals as desired because of the increased accuracy that can be gained by allotting more females to each male under test.

The information from a progeny test can be used in the calculation of the "equal-parent index" (sometimes referred to as the "midparent index"). If the progeny receives a sample half of each of its parents' genotypes and the plus and minus effects of Mendelian errors and errors of appraisal tend to cancel each other in averages of the progeny and dams, then Average of progeny = sire/2 + (average of dams)/2 or

$$\text{Sire} = 2(\text{average of progeny}) - (\text{average of dams})$$

Mating Methods

Once the selected individuals have been chosen, they may be mated in various ways. The process known as "breeding" includes the judicious selection and mating of individuals for particular purposes.

Random Mating (Panmixis)

If the breeder places no mating restraints upon the selected individuals, their gametes are likely to randomly unite by chance alone. This is commonly the case with **outcrossing** (non-self-fertilizing) plants. Wind or insects carry pollen from one plant to another in essentially a random number. Even livestock such as sheep and range cattle are usually bred panmicticly. The males locate females as they come into heat, copulate with ("cover") and inseminate them without any artificial restrictions as they forage for food over large tracts of grazing land. Most of the food that reaches our table is produced by random mating because it is the most economical mating method; relatively little manual labor is expended by the shepherd or herdsman other than keeping the flock or herd together, warding off predators, etc.

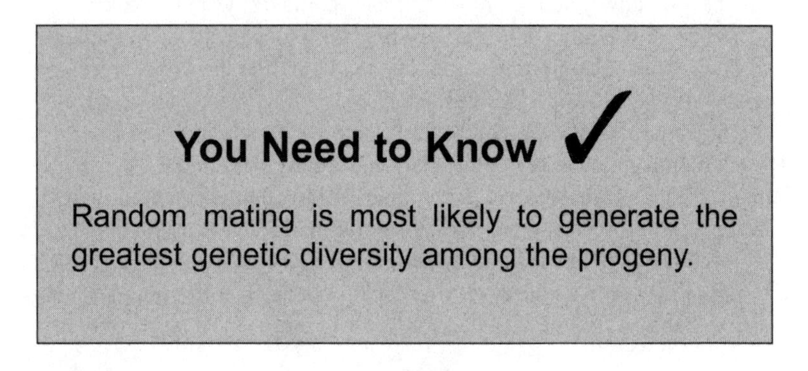

Positive Assortative Mating

This method involves mating individuals that are more alike, either phenotypically or genotypically, than the average of the selected group.

Based on Genetic Relatedness. Inbreeding is the mating of individuals more closely related than the average of the population to which they belong. Figure 9-7(a) shows a pedigree in which no inbreeding is evident because there is no common ancestral pathway from B to C (D, E, F, and G all being unrelated). In the inbred pedigree of Figure 9-7(b), B and C have the same parents and thus are full sibs (brothers and/or sisters). In the standard pedigree form shown in Figure 9-7(b), sires appear on the upper lines and dams on the lower lines. Thus B and D are males; C and E are females. It is desirable to convert a standard pedigree into an arrow diagram for analysis [Figure 9-7(c)]. The **coefficient of relationship** (R) estimates the percentage of genes held in common by two individuals because of their common ancestry. Since one transmits only a sample half of one's genotype to one's offspring, each arrow in the diagram represents a probability of H. The sum (Σ) of all pathways between two individuals through common ancestors is the coefficient of relationship.

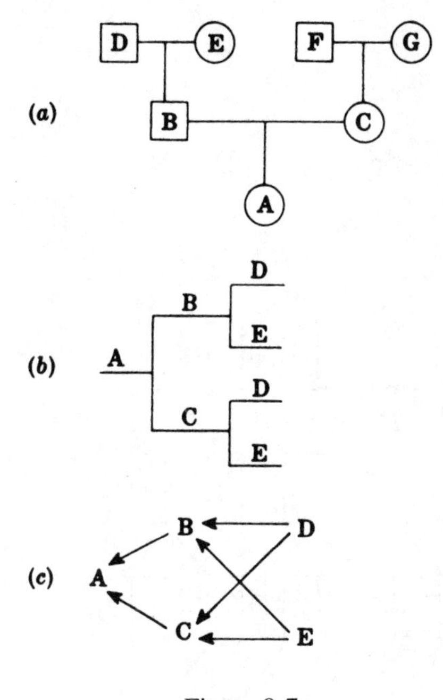

Figure 9-7

When matings occur only between closely related individuals (inbreeding), the genetic effect is an increase in homozygosity. The most intense form of inbreeding is self-fertilization. If we start with a population containing 100 heterozygotes individuals (*Aa*) as shown in Table 9.1, the expected number of homozygous genotypes is increased by 50 percent due to selfing in each generation.

Table 9.1

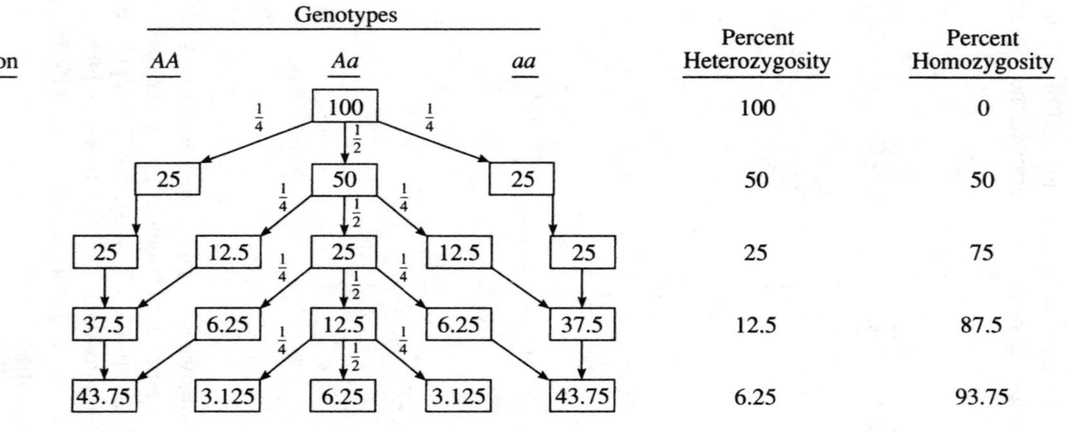

Generation	AA		Aa		aa	Percent Heterozygosity	Percent Homozygosity
0			100			100	0
1	25		50		25	50	50
2	25	12.5	25	12.5	25	25	75
3	37.5	6.25	12.5	6.25	37.5	12.5	87.5
4	43.75	3.125	6.25	3.125	43.75	6.25	93.75

Other less intense forms of inbreeding produce a less rapid approach to homozygosity, shown graphically in Figure 9-8. As homozygosity increases in a population, due to either inbreeding or selection, the genetic variability of the population decreases. Since heritability depends upon the relative amount of genetic variability, it also decreases so that in the limiting case (pure line) heritability becomes zero.

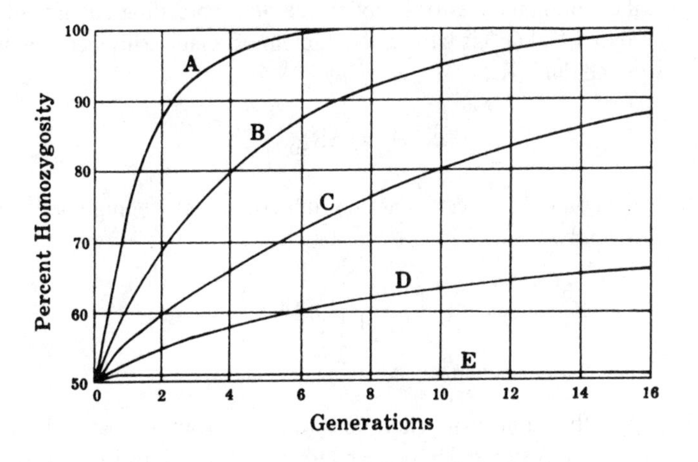

Figure 9-8

When population size is reduced to a small isolated unit containing less than about 50 individuals, inbreeding very likely will result in a detectable increase in genetic uniformity. The **coefficient of inbreeding** (symbolized by F) is a useful indicator of the probable effect that inbreeding has had at two levels.

(1) On an *individual basis*, the coefficient of inbreeding indicates the probability that the two alleles at any locus are identical by descent, i.e., they are both replication products of a gene present in a common ancestor.
(2) On a *population basis*, the coefficient of inbreeding indicates the percentage of all loci that were heterozygous in the base population that now have probably become homozygous due to the effects of inbreeding. The base population is that point in history of the pop-

ulation from which we desire to begin a calculation of the effects of inbreeding. Many loci are probably homozygous at the time we establish our base population. The inbreeding coefficient then measures the *additional* increase in homozygosity due to matings between closely related individuals.

The coefficient of inbreeding (F) can be determined for an individual in a pedigree by several similar methods.

(1) If the common ancestor is not inbred, the inbreeding coefficient of an individual (F_x) is half the coefficient of relationship between the sire and dam (R_{SD}):

$$F_x = \tfrac{1}{2} R_{SD}$$

(2) If the common ancestors are not inbred, the inbreeding coefficient is given by

$$F_x = \Sigma \left(\frac{1}{2} \right)^{p_1 + p_2 + 1}$$

where p_1 is the number of generations (arrows) from one parent back to the common ancestor and p_2 is the number of generations from the other parent back to the same ancestor.

(3) If the common ancestors are inbred (F_A), the inbreeding coefficient of the individual must be corrected for this factor:

$$F_x = \Sigma \left[\left(\frac{1}{2} \right)^{p_1 + p_2 + 1} (1 + F_A) \right]$$

(4) The coefficient of inbreeding of an individual may be calculated by counting the number of arrows (n) that connect the individual through one parent back to the common ancestor and back again to the other parent, and applying the formula

$$F_x = \Sigma(\tfrac{1}{2})^n(1 + F_A)$$

The following table will be helpful in calculating F.

n	1	2	3	4	5	6	7	8	9
$(\frac{1}{2})^n$	0.5000	0.2500	0.1250	0.0625	0.0312	0.0156	0.0078	0.0039	0.0019

Linebreeding is a special form of inbreeding utilized for the purpose of maintaining a high genetic relationship to a desirable ancestor. Figure 9-9 shows a pedigree in which close linebreeding to B has been practiced so that A possesses more than 50 percent of B's genes. D possesses 50 percent of B's genes and transmits 25 percent to C. B also contributes 50 percent of his genes to C. Hence, C contains 50 percent + 25 percent = 75 percent B genes and transmits half of them (37.5 percent) to A. B also contributes 50 percent of his genes to A. Therefore, A has 50 percent + 37.5 percent = 87.5 percent of B's genes.

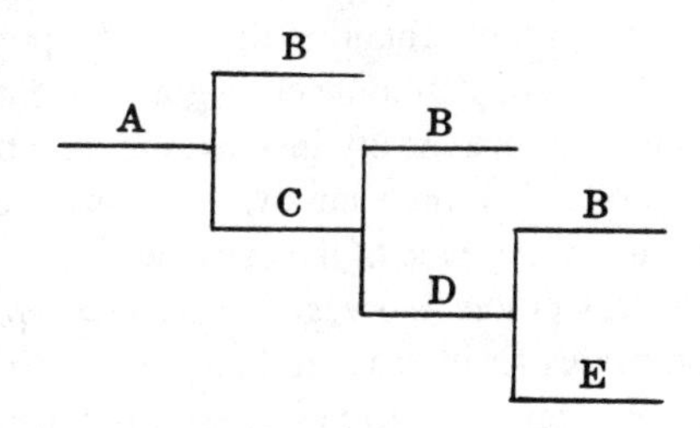

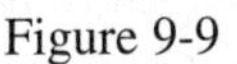
Figure 9-9

Based on Phenotypic Similarity. Positive phenotypic assortative mating is seldom practiced in its purest form among the selected individuals, i.e., mating only "look-alikes" or those with nearly the same selection indices. However, it can be used in conjunction with random mating; a few of the best among the selected group are "hand-coupled," artificially cross-pollinated, or otherwise forced to breed. Both inbreeding and positive phenotypic assortative mating tend to reduce genetic heterozygosity but the theoretical end results are quite different. The rate at which heterozygous loci can be fixed (brought to homozygosity) in a population can be greatly accelerated by combining a system of close inbreeding with the additional restriction of positive phenotypic assortative mating; in other words, they must also "look" alike.

Negative Assortative Mating

Based on Genetic Relatedness. When a mating involves individuals that are more distantly related than the average of the selected group, it is classified as a negative genetic assortative mating. This may involve crossing individuals belonging to different families or crossing different inbred varieties of plants or crossing different breeds of livestock. It may occasionally involve crossing closely related species such as the horse and ass (donkey, burro) to produce the hybrid male. The usual purpose of these "outcrosses" is an attempt to produce offspring of superior phenotypic quality (but not necessarily inbreeding value) to that normally found in the parental populations.

Many recessives remain hidden in heterozygous conditions in noninbred populations, but as homozygosity increases in an inbred population, there is a greater probability that recessive traits, many of which are deleterious, will begin to appear. One of the consequences of inbreeding is a loss in vigor (i.e., less productive vegetatively and reproductively) that commonly accompanies an increase in homozygosity (**inbreeding depression**). Crosses between inbred lines usually produce a vigorous hybrid F_1 generation. This increased "fitness" of heterozygous individuals has been termed **heterosis**. The genetic basis of heterosis is still a subject of controversy, largely centered about two theories.

(1) The *dominance theory* of heterosis. Hybrid vigor is presumed to result from the action and interaction of dominant growth or fitness factors.
(2) The *overdominance theory* of heterosis. Heterozygosity per se is assumed to produce hybrid vigor.

Phenotypic variability in the hybrid generation is generally much less than that exhibited by the inbred parental lines (Figure 9-10). This indicated that the heterozygotes are less subject to environmental influences than the homozygotes. Geneticists use the term "buffering" to indicate that the organism's development is highly regulated genetically ("canalized"). Another term often used in this connection is **homeostasis**, which signifies the maintenance of a "steady state" in the development and physiology of the organism within the normal range of environmental fluctuations.

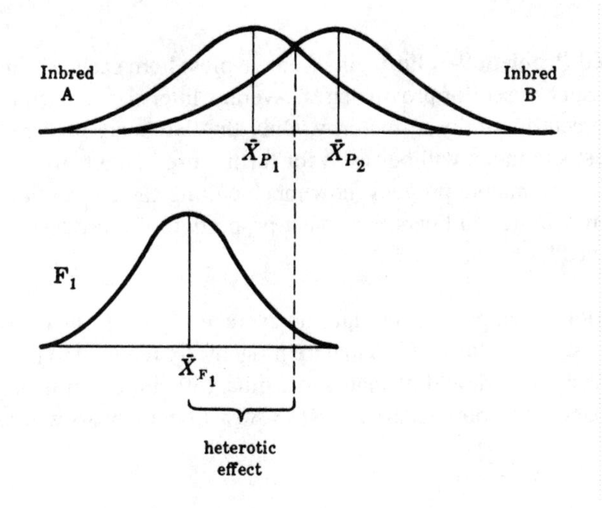

Figure 9-10

A rough guide to the estimation of heterotic effects (H) is obtained by noting the average excess in vigor that F_1 hybrids exhibit over the midpoint between the means of the inbred parental lines (Figure 9-10).

$$H_{F_1} = \overline{X}_{F_1} - \frac{1}{2}\left(\overline{X}_{P_1} + \overline{X}_{P_2}\right)$$

The heterosis exhibited by an F_2 population is commonly observed to be half of that manifested by the F_1 hybrids.

Based on Phenotypic Dissimilarity. When intermediate phenotypes are preferred, they are more likely to be produced by mating opposite phenotypes. For example, general-purpose cattle can be produced by crossing a beef type with a dairy type. The offspring commonly produce an intermediate yield of milk and hang up a fair carcass when slaughtered (although generally not as good in either respect as the parental types). The same is true of the offspring from crossing an egg type (such as the Leghorn breed of chicken) with a meat type (such as the Cornish). Crossing phenotypic opposites may also be made to correct specific defects.

Solved Problem 9-1. Fifty gilts (female pigs) born each year in a given herd can be used for proving sires. Average litter size at birth is 10 with 10 percent mortality to maturity. Only the five boars (males) with the highest sire index will be saved for further use in the herd. If each test requires 18 mature progeny, how much culling can be practiced among the progeny-tested boars, i.e., what proportion of those tested will not be saved?

Solution. Each gilt will produce an average of $10 - (0.1)(10) = 9$ progeny raised to maturity. If 18 mature progeny are required to prove a sire, then each boar should be mated to 2 gilts. (50 gilts)/(2 gilts per boar) = 25 boars can be proved. $20/25 = 4/5 = 80\%$ of these boars will be culled.

Chapter 10
POPULATION GENETICS

IN THIS CHAPTER:

- ✔ *Hardy-Weinberg Equilibrium*
- ✔ *Calculating Gene Frequencies*
- ✔ *Testing a Locus for Equilibrium*

Hardy-Weinberg Equilibrium

A **Mendelian population** may be considered to be a group of sexually reproducing organisms with a relatively close degree of genetic relationship (such as a species, subspecies, breed, variety, strain) residing within defined geographic boundaries wherein interbreeding occurs. If all the gametes produced by a Mendelian population are considered as a hypothetical mixture of genetic units from which the next generation will arise, we have the concept of a **gene pool**.

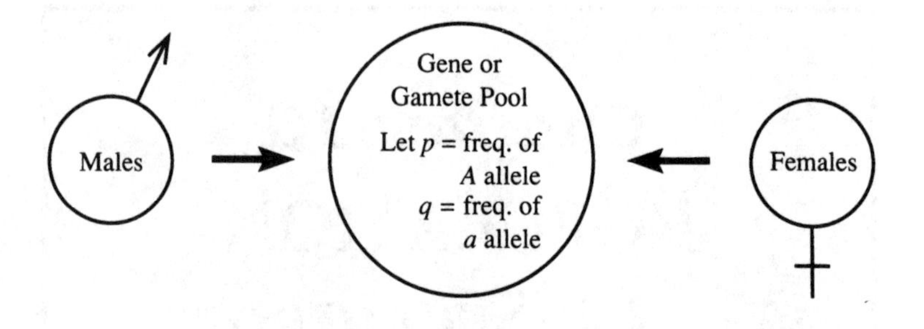

If we consider a pair of alleles (*A* and *a*), we will find that the percentage of gametes in the gene pool bearing *A* or *a* will depend upon the genotypic frequencies of the parental generation whose gametes form the pool. For example, if most of the population were of the recessive genotype *aa*, then the frequency of the recessive allele in the gene pool would be relatively high, and the percentage of gametes bearing the dominant (*A*) allele would be correspondingly low.

When matings between members of a population are completely at random, i.e., when every male gamete in the gene pool has an equal opportunity of uniting with every female gamete, then the zygotic frequencies expected in the next generation may be predicted from a knowledge of the gene (allelic) frequencies in the gene pool of the parental population. That is, given the relative frequencies of *A* and *a* gametes in the gene pool, we can calculate (on the basis of the chance union of gametes) the expected frequencies of progeny genotypes and phenotypes. If p = percentage of *A* alleles in the gene pool and q = percentage of *a* alleles, then we can use the checkerboard method to produce all the possible chance combination of these gametes.

♀ \ ♂	p (A)	q (a)
p (A)	p^2 AA	pq Aa
q (a)	pq Aa	q^2 aa

Note that $p + q = 1$, i.e., the percentage of A and a gametes must add to 100 percent in order to account for all of the gametes in the gene pool. The expected genotypic (zygotic) frequencies in the next generation then may be summarized as follows:

$$(p + q)^2 = \underset{AA}{p^2} + \underset{Aa}{2pq} + \underset{aa}{q^2} = 1.0$$

Thus, p^2 is the fraction of the next generation expected to be homozygous dominant (AA), $2pq$ is the fraction expected to be heterozygous (Aa), and q^2 is the fraction expected to be recessive (aa). All of these genotypic fractions must add to unity to account for all genotypes in the progeny population.

This formula, expressing the genotypic expectations of progeny in terms of the gametic (allelic) frequencies of the parental gene pool, is called the **Hardy-Weinberg law**. If a population conforms to the conditions on which this formula is based, there should be no change in the gametic or the zygotic frequencies from generation to generation. Should a population initially be in disequilibrium, one generation of random mating is sufficient to bring it into genetic equilibrium and thereafter the population will remain in equilibrium (unchanging in gametic and zygotic frequencies) as long as the Hardy-Weinberg conditions persist.

Several assumptions underlie the attainment of genetic equilibrium as expressed in the Hardy-Weinberg equation.

(1) The population is infinitely large and mates at random (**panmictic**).
(2) No selection is operative, i.e., each genotype under consideration can survive just as well as any other (no differential mortality), and each genotype is equally efficient in the production of progeny (no differential reproduction).
(3) The population is closed, i.e., no immigration of individuals from another population into nor emigration from the population under consideration is allowed.
(4) There is no mutation from one allelic state to another. Mutation may be allowed if the forward and back mutation rates are equivalent, i.e., A mutates to a with the same frequency that a mutates to A.

(5) Meiosis is normal so that chance is the only factor operative in gametogenesis.

If we define **evolution** as any change in a population from the Hardy-Weinberg equilibrium conditions, then a violation of one or more of the Hardy-Weinberg restrictions could cause the population to move away from the gametic and zygotic equilibrium frequencies. Changes in gene frequencies can be produced by a reduction in population size; by selection, migration, or mutation pressures; or by **meiotic drive** (non-random assortment of chromosomes). No population is infinitely large, spontaneous mutations cannot be prevented, selection and migration pressures usually exist in most natural populations, etc., so it may be surprising to learn that despite these violations of Hardy-Weinberg restrictions many genes do conform within statistically acceptable limits, to equilibrium conditions between two successive generations.

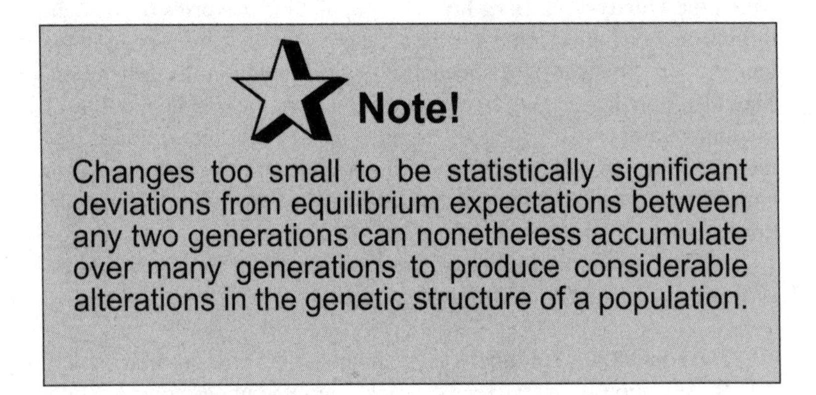

A **race** is a genetically (and usually geographically) distinctive interbreeding population of a species. The number of races one wishes to recognize generally depends on the purpose of the investigation. Populations that differ significantly in gene frequencies at one or more loci may be considered as different races. Human races are defined on the basis of gene frequency differences in qualitative traits such as blood groups, hair texture, eye color, etc., as well as by mean and standard deviation differences in quantitative traits such as skin color, body build, shapes of noses, lips, eyes, etc. Races of a given species can freely interbreed with one another. Members of different species, however, are reproductively isolated to a recognizable degree.

Subspecies are races that have been given distinctive taxonomic names. Varieties, breeds, strains, etc., of cultivated plants or domesticated animals may also be equated with the racial concept. Geographic isolation is usually required for populations of a species to become distinctive races. Race formation is a prerequisite to the splitting of one species into two or more species (**speciation**). Differentiation at many loci over many generations is generally required to reproductively isolate these groups by time of breeding, behavioral differences, ecological requirements, hybrid inviability, hybrid sterility, and other such mechanisms.

Equilibrium at an autosomal genetic locus becomes fully established in a nonequilibrium population after one generation of random mating under Hardy-Weinberg conditions regardless of the number of alleles at that locus. However, when autosomal allelic frequencies are dissimilar in the sexes, they become equilibrated after one generation of random mating, but the genotypic frequencies do not become equilibrated until the second generation of random mating. If the frequencies of sex-linked alleles are unequal in the sexes, the equilibrium values are approached rapidly during successive generations of random mating in an oscillatory manner by the two sexes. This phenomenon derives from the fact that females (XX) carry twice as many sex-linked alleles as do males (XY). Females receive their sex-linked heredity equally from both parents, but males receive their sex-linked heredity only from their mothers. The difference between the allelic frequencies in males and females is halved in each generation under random mating. Within each sex, the derivation from equilibrium is halved in each generation, with sign reversed. The average frequency of one allele ($\overline{p}$) in the entire population is also the equilibrium approached by each sex during successive generations of random mating.

$$\overline{p} = \frac{2}{3} p_f + \frac{1}{3} p_m$$

Although alleles at a single autosomal locus reach equilibrium following one generation of random mating, gametic equilibrium involving two independently assorting genes is approached rapidly over a number of generations. At equilibrium, the product of coupling gametes equals the product of repulsion gametes.

Example 10.1 Consider one locus with alleles *A* and *a* at frequencies represented by *p* and *q*, respectively. A second locus has alleles *B* and *b* at frequencies *r* and *s*, respectively. The expected frequencies of coupling gametes *AB* and *ab* are *pr* and *qs*, respectively. The expected frequencies of repulsion gametes *Ab* and *aB* are *ps* and *qr*, respectively. At equilibrium, $(pr)(qs) = (ps)(qr)$. Also at equilibrium, the disequilibrium coefficient (d) is $d = (pr)(qs) - (ps)(qr) = 0$.

For independently assorting loci under random mating, the disequilibrium value of *d* is halved in each generation during the approach to equilibrium by linked genes, however, is slowed by comparison because they recombine less frequently than unlinked genes (i.e., less than 50 percent recombination). The closer the linkage, the longer it takes to reach equilibrium. The disequilibrium (d_t) that exists at any generation (t) is expressed as:

$$d_t = (1 - \mathrm{r})\, d_{t-1}$$

where r = frequency of recombination and d_{t-1} = disequilibrium in the previous generation.

Example 10.2 If $d = 0.25$ initially and the 2 loci experience 20 percent recombination (i.e., the loci are 20 map units apart), the disequilibrium that would be expected after one generation of random mating is $d_t = (1 - 0.2)(0.25) = 0.2$. This represents $0.20/0.25 = 0.8$ or 80 percent of the maximum disequilibrium that could exist for a pair of linked loci.

Calculating Gene Frequencies

Autosomal Loci with Two Alleles

Codominant Autosomal Alleles. When codominant alleles are present in a two-allele system, each genotype has a distinctive phenotype. The numbers of each allele in both homozygous and heterozygous conditions may be counted in a sample of individuals from the population and expressed as a percentage of the total number of alleles in the sample. If the sample is representative of the entire population (containing proportionately the same numbers of genotypes as found in the entire population), then we can obtain an estimate of the allelic frequencies in the gene pool. Given a sample of N individuals of which D are homozygous for one allele (A^1A^1), H are heterozygous (A^1A^2), and R are homozygous for the other allele (A^2A^2), then N = D + H + R. Since each of the N individuals are diploid at this locus, there are 2N alleles represented in the sample. Each A^1A^1 genotype has two A^1 alleles. Heterozygotes have only one A^1 allele. Letting p represent the frequency of the A^1 allele and q the frequency of the A^2 allele, we have

$$p = \frac{2D+H}{2N} = \frac{D+\frac{1}{2}H}{N} \qquad q = \frac{H+2R}{2N} = \frac{\frac{1}{2}H+R}{N}$$

Dominant and Recessive Autosomal Alleles. Determining the gene frequencies for alleles that exhibit dominance and recessive relationships requires a different approach from that used with codominant alleles. A dominant phenotype may have either of two genotypes, *AA* or *Aa*, but we have no way (other than by laboriously testcrossing each dominant phenotype) of distinguishing how many are homozygous or heterozygous in our sample. The only phenotype whose genotype is known for certain is the recessive allele (*aa*). If the population is in equilibrium, then we can obtain an estimate of q (the frequency of the recessive allele) from q^2 (the frequency of the recessive genotype or phenotype).

Example 10.3. If 75 percent of a population was of the dominant phenotype (A–), then 25 percent would have the recessive phenotype (*aa*). If the population is in equilibrium with respect to this locus, we expect q2 = frequency of aa. Then q2 = 0.25, q = 0.5, p = 1 – q = 0.5.

Sex-Influenced Traits. The expression of dominance and recessive relationships may be markedly changed in some genes when exposed to different environmental conditions, most notable of which are the sex hormones. In sex-influenced traits, the heterozygous genotype usually will produce different phenotypes in the two sexes, making the dominance and recessive relationships of the alleles appear to reverse themselves. Determination of allelic frequencies must be indirectly made in one sex by taking the square root of the frequency of the recessive phenotype ($q = \sqrt{q^2}$). A similar approach in the opposite sex should give an estimate of p. Corroboration of sex influence is obtained if these estimates of p and q made in different sexes add close to unity.

Autosomal Loci with Multiple Alleles

If we consider three alleles, A, a', and a, with the dominance hierarchy $A > a' > a$, occurring in the gene pool with respective frequencies p, q, and r, then random mating will generate zygotes with the following frequencies:

$$(p + q + r)^2 = p^2 + 2pq + 2pr + q^2 + 2qr + r^2 = 1$$

$$\begin{array}{llllllll} \text{Genotypes:} & \underbrace{AA \quad Aa' \quad Aa} & & & \underbrace{a'a' \quad a'a} & & \underbrace{aa} \\ \text{Phenotypes:} & A & & & a' & & a \end{array}$$

For ease in calculation of a given allelic frequency, it may be possible to group the phenotypes of the population into just two types.

Example 10.4 In a multiple allelic system where $A > a' > a$, we could calculate the frequency of the top dominant allele A by considering the dominant phenotype (A) in contrast to all other phenotypes produced by alleles at this locus. The latter group may be considered to be produced by an allele a_x which is recessive to A. Let p = frequency of allele A,

q = frequency of allele a_x, q^2 = frequency of phenotypes other than A. Given this, $p = 1 - q$ = frequency of gene A.

Many multiple allelic series involve codominant relationships such as $(A^1 = A^2) > a$, with respective frequencies p, q, and r. More genotypes can be phenotypically recognized in codominant systems than in systems without codominance.

$$(p + q + r)^2 = p^2 + 2pr + 2pq + q^2 + 2qr + r^2 = 1$$

Genotypes:	A^1A^1	A^1a	A^1A^2	A^2A^2	A^2a	aa
Phenotypes:	A^1		A^1A^2	A^2		a

Sex-Linked Loci

Codominant Sex-Linked Alleles. Data from both males and females can be used in the direct computation of sex-linked codominant allelic frequencies. Bear in mind that in organisms with an X-Y mechanism of sex determination, the heterozygous condition can only appear in females. Males are hemizygous for sex-linked genes.

Example 10.5 In domestic cats, black melanin pigment is deposited in the hair by a sex-linked gene; its alternative allele produces yellow hair. Random inactivation of one of the X chromosomes occurs in each cell of female embryos. Heterozygous females are thus genetic mosaics, having patches of all-black and all-yellow hairs called tortoise-shell pattern. Since only one sex-linked allele is active in any cell, the inheritance is not really codominant, but the genetic symbolism used is the same as that for codominant alleles.

	Phenotypes		
	Black	Tortoise-Shell	Yellow
Females	C^bC^b	C^bC^y	C^yC^y
Males	C^bY	—	C^yY

Let p = frequency of C^b, q = frequency of C^y.

$$p = \frac{2\begin{pmatrix}\text{no. of black}\\ \text{females}\end{pmatrix} + \begin{pmatrix}\text{no. of tortoise-}\\ \text{shell females}\end{pmatrix} + \begin{pmatrix}\text{no. of black}\\ \text{males}\end{pmatrix}}{2(\text{no. females}) + \text{no. males}}$$

$$q = \frac{2\begin{pmatrix}\text{no. of yellow}\\ \text{females}\end{pmatrix} + \begin{pmatrix}\text{no. of tortoise-}\\ \text{shell females}\end{pmatrix} + \begin{pmatrix}\text{no. of yellow}\\ \text{males}\end{pmatrix}}{2(\text{no. females}) + \text{no. males}}$$

Dominant and Recessive Sex-Linked Alleles. Since each male possesses only one sex-linked allele, the frequency of a sex-linked trait among males is a direct measure of the allelic frequency in the population, assuming, of course, that the allelic frequencies thus determined are representative of the allelic frequencies among females as well.

Testing a Locus for Equilibrium

In cases where dominance is involved, the heterozygous class is indistinguishable phenotypically from the homozygous dominant class. Hence there is no way of checking the Hardy-Weinberg expectations against observed sample data unless the dominant phenotypes have been genetically analyzed by observation of their progeny from testcrosses. Only when codominant alleles are involved can we easily check our observations against the expected equilibrium values through the chi-square test.

Degrees of Freedom

The number of variables in chi-square tests of Hardy-Weinberg equilibrium is not simply the number of phenotypes minus 1 (as in chi-square tests of classical Mendelian ratios). The number of observed variables (number of phenotypes = k) is further restricted by testing their conformity to an expected Hardy-Weinberg frequency ratio generated by a number of additional variables (number of alleles, or allelic frequencies = r). We have $(k - 1)$ degrees of freedom in the number of phenotypes, $(r - 1)$ degrees of freedom in establishing the frequencies for the r alleles. The combined number of degrees of freedom is $(k - 1) - (r - 1) = k - r$. Even in most chi-square tests for equilibrium involving multiple alleles, the number of degrees of freedom is the number of phenotypes minus the number of alleles.

Solved Problem 10.1. In a population gene pool, the alleles *A* and *a* are at initial frequencies p and q, respectively. Prove that the gene frequencies and the zygotic frequencies do not change from generation to generation as long as the Hardy-Weinberg conditions are maintained.

Solution.
Zygotic frequencies generated by random mating are:

$$p^2(AA) + 2pq(Aa) + q^2(aa) = 1$$

All of the gametes of *AA* individuals and half of the gametes of heterozygotes will bear the dominant allele (*A*). Then the frequency of *A* in the gene pool of the next generation is:

$$p^2 + pq = p^2 + p(1 - p) = p^2 + p - p^2 = p$$

Thus, each generation of random mating under Hardy-Weinberg conditions fails to change either the allele or zygotic frequencies.

Index